Advances in Ceramic Armor IV

Advances in Ceramic Armor IV

A Collection of Papers Presented at the 32nd International Conference on Advanced Ceramics and Composites
January 27–February 1, 2008
Daytona Beach, Florida

Editor
Lisa Prokurat Franks

Volume Editors
Tatsuki Ohji
Andrew Wereszczak

A John Wiley & Sons, Inc., Publication

Published by John Wiley & Sons, Inc., Hoboken, New Jersey.
Published simultaneously in Canada.

Library of Congress Cataloging-in-Publication Data is available.

ISBN: 978-0-470-34497-2

Contents

OPAQUE CERAMICS

NOVEL EVALUATION AND CHARACTERIZATION

Preface

The 32nd International Conference and Exposition on Advanced Ceramics & Composites, January 27- February 1, 2008, in Daytona Beach, Florida marks the sixth consecutive year for modelers, experimentalists, processors, testers, fabricators, manufacturers, managers, and simply ceramists to present and discuss the latest issues in ceramic armor. This annual meeting is organized and sponsored by The American Ceramic Society (ACerS) and the ACerS Engineering Ceramics Division, and is affectionately known as the Cocoa Beach conference because it has been held in Cocoa Beach, Florida since its inception. Although there was concern about the risk of a change in location of such a time honored tradition as the Cocoa Beach conference, the move up the coast to Daytona Beach has proved successful for the Ceramic Armor Symposium. Currently, there is no other unclassified forum in government, industry, or academia that brings together these interdependent disciplines to address the challenges in ceramic armor, and consequently, we needed a more spacious venue.

For each of the past six years, we have been privileged to have technical experts who are not only enthusiastic about their work, but also willing to address the tough or controversial aspects in lively discussion. This meeting continued the tradition of an unclassified forum on an unprecedented international scale.

Although special thanks must go to the Organizing Committee for its exceptional efforts to support the Ceramic Armor Symposium, I want to especially extend my sincere thanks to the participants. With your reasonable and concise feedback we have been able to keep the successes both technically and logistically, and make improvements where needed. Since the first Ceramic Armor session, many of you have ensured that the "Cocoa Beach" conference, or now fondly "Cocotona", is your must-attend-event, and the core group of participants has continued to grow.

Thank you colleagues and friends.

LISA PROKURAT FRANKS
U.S. Army TARDEC

Introduction

Organized by the Engineering Ceramics Division (ECD) in conjunction with the Basic Science Division (BSD) of The American Ceramic Society (ACerS), the 32nd International Conference on Advanced Ceramics and Composites (ICACC) was held on January 27 to February 1, 2008, in Daytona Beach, Florida. 2008 was the second year that the meeting venue changed from Cocoa Beach, where ICACC was originated in January 1977 and was fostered to establish a meeting that is today the most preeminent international conference on advanced ceramics and composites

The 32nd ICACC hosted 1,247 attendees from 40 countries and 724 presentations on topics ranging from ceramic nanomaterials to structural reliability of ceramic components, demonstrating the linkage between materials science developments at the atomic level and macro level structural applications. The conference was organized into the following symposia and focused sessions:

Symposium 1	Mechanical Behavior and Structural Design of Monolithic and Composite Ceramics
Symposium 2	Advanced Ceramic Coatings for Structural, Environmental, and Functional Applications
Symposium 3	5th International Symposium on Solid Oxide Fuel Cells (SOFC): Materials, Science, and Technology
Symposium 4	Ceramic Armor
Symposium 5	Next Generation Bioceramics
Symposium 6	2nd International Symposium on Thermoelectric Materials for Power Conversion Applications
Symposium 7	2nd International Symposium on Nanostructured Materials and Nanotechnology: Development and Applications
Symposium 8	Advanced Processing & Manufacturing Technologies for Structural & Multifunctional Materials and Systems (APMT): An International Symposium in Honor of Prof. Yoshinari Miyamoto
Symposium 9	Porous Ceramics: Novel Developments and Applications

Symposium 10	Basic Science of Multifunctional Ceramics
Symposium 11	Science of Ceramic Interfaces: An International Symposium Memorializing Dr. Rowland M. Cannon
Focused Session 1	Geopolymers
Focused Session 2	Materials for Solid State Lighting

Peer reviewed papers were divided into nine issues of the 2008 Ceramic Engineering & Science Proceedings (CESP); Volume 29, Issues 2-10, as outlined below:

- Mechanical Properties and Processing of Ceramic Binary, Ternary and Composite Systems, Vol. 29, Is 2 (includes papers from symposium 1)
- Corrosion, Wear, Fatigue, and Reliability of Ceramics, Vol. 29, Is 3 (includes papers from symposium 1)
- Advanced Ceramic Coatings and Interfaces III, Vol. 29, Is 4 (includes papers from symposium 2)
- Advances in Solid Oxide Fuel Cells IV, Vol. 29, Is 5 (includes papers from symposium 3)
- Advances in Ceramic Armor IV, Vol. 29, Is 6 (includes papers from symposium 4)
- Advances in Bioceramics and Porous Ceramics, Vol. 29, Is 7 (includes papers from symposia 5 and 9)
- Nanostructured Materials and Nanotechnology II, Vol. 29, Is 8 (includes papers from symposium 7)
- Advanced Processing and Manufacturing Technologies for Structural and Multifunctional Materials II, Vol. 29, Is 9 (includes papers from symposium 8)
- Developments in Strategic Materials, Vol. 29, Is 10 (includes papers from symposia 6, 10, and 11, and focused sessions 1 and 2)

The organization of the Daytona Beach meeting and the publication of these proceedings were possible thanks to the professional staff of ACerS and the tireless dedication of many ECD and BSD members. We would especially like to express our sincere thanks to the symposia organizers, session chairs, presenters and conference attendees, for their efforts and enthusiastic participation in the vibrant and cutting-edge conference.

ACerS and the ECD invite you to attend the 33rd International Conference on Advanced Ceramics and Composites (http://www.ceramics.org/daytona2009) January 18–23, 2009 in Daytona Beach, Florida.

TATSUKI OHJI and ANDREW A. WERESZCZAK, Volume Editors
July 2008

Transparent Glasses and Ceramics

MESOMECHANICAL CONSTITUTIVE RELATIONS FOR GLASS AND CERAMIC ARMOR

D. R. Curran, D. A. Shockey, and J. W. Simons
SRI International
Menlo Park, CA 94025

ABSTRACT

A major challenge in achieving a physics-based computational capability for designing glass and ceramic armor is a damage evolution and fragment flow model that is usable in continuum codes. We describe a model that uses microfailure and fragment flow constitutive data, show how the model links to continuum models, and compare computational results with glass penetration tests.

BACKGROUND AND PROGRAM GOALS

Improved mesomechanical constitutive relations for glass targets undergoing microscopic damage are needed for efficient design of transparent armor. The role of the mesomechanical models is to relate material failure on the microscopic level to continuum behavior, and to give guidance to continuum models that are used in hydrocodes.

Penetration of thick targets of both ductile and brittle materials occurs by the formation of a region of yielded, flowing material at the penetrator–target interface. The flow of this material allows penetration to occur. For brittle materials like glass and ceramics, the yielded material is observed to consist of fine fragments in a thin region called the Mescall zone (MZ)[1].

Our goal is to construct a mesomodel that describes the microdamage evolution, i.e. the nucleation, growth, and coalescence of microcracks to form the MZ, and subsequent granular flow of the comminuted material out of the path of the advancing penetrator. Our mesomodel is empirical, and is based on observations and data from experiments designed to measure microdamage evolution and fragment flow.

In this paper we describe a proposed initial framework for a mesomodel, incorporate available data, discuss preliminary correlations of predictions with observations, and discuss future proposed experiments.

MESOMECHANICAL APPROACH

Empirically-based mesomechanical constitutive relations have been successfully developed during the past several decades to relate material failure in metals and composites to the underlying microscopic processes, thereby helping to select appropriate continuum models and resolve apparent paradoxes[1-4]. The key to this approach has been the development of experiments for determination of "nucleation and growth to fragmentation" (NAG/FRAG) laws in a relevant volume element (RVE) for the evolution of size distributions of microscopic voids and cracks and their coalescence to form fragments, as well as the subsequent motion of these fragments.

NAG/FRAG experiments are designed to measure key properties. As summarized in the 2004 book by Kanel, Razorenov, and Fortov[5], glass presents several challenges, as follows.

- Flaw sites: Whereas most brittle materials contain internal flaws that can serve as microcrack nucleation sites, high quality glass has primarily only surface flaws. This means that in uniaxial strain plate impact experiments, for example, the microdamage should be localized in a region adjacent to the impact surface.
- HEL: Glasses do not exhibit a distinct Hugoniot elastic limit in plate impact experiments, partly because of a convex downward curvature of the Hugoniot at low pressures.

Molecular structure: At high pressures, brittle glasses become ductile. The molecular structure of glass allows densification without cracking at pressures exceeding 7-10 GPa. At such pressures, evidence of yield may vanish, and the response may be difficult to distinguish between elastic and hydrostatic.

Prior experiments have emphasized instrumented long and short rod penetration tests ranging from near the dwell transition to steady-state penetration. The measurements include x-ray or optical "snapshots" of the position of the macrocrack front, MZ front, and penetrator tail position[6,7]. However, until recently, experiments specifically designed to yield NAG/FRAG relations for the evolution of microscopic damage were lacking. In the present paper we focus on three types of experiments that can potentially provide such information:

1. Plate impact (uniaxial strain) experiments. These experiments simulate the loading conditions on-axis under the penetrator nose during the impact shock response.
2. Partial penetration of non-eroding rods. These experiments reproduce the loading conditions near the nose of elastic penetrators at penetration rates of several hundred m/s.
3. Quasistatic material property tests, including compression-torsion tests of powders. These experiments provide basic properties of the MZ material.

To guide planning and interpretation of the above experiments, we start with a conceptual mesomodel (CMM), which will serve as an initial framework to be modified as we obtain more microdamage evolution data.

CONCEPTUAL MESOMECHANICAL MODEL

The CMM is based on modifications of the FRAGBED2 (FB2) mesomodel of non-elastic flow in brittle materials[8]. The non-elastic flow is assumed to be totally due to elastic fragments sliding frictionally on inter-fragment interfaces, and is treated by analogy to multi-plane plasticity models based on atomic dislocation dynamics, i.e. we focus on the movement of lines of holes between the fragments, called macrodislocations (MDs), on a finite number of slip planes. The flow is inherently rate-dependent because of finite crack nucleation and growth rates and fragment inertia. Linear Elastic Fracture Mechanics (LEFM) is assumed to govern the microcrack nucleation, and the "fracture toughness" is a property that represents the material's brittleness.

Micrographs of fragmented ceramics suggest that the fragmented bed is initially a jumbled array of different-sized fragments in which flow is inhibited because the fragments block each other, and the associated MDs are "pinned". The "yield" condition in the CMM is thus an "unpinning" condition. Figure 1 is a schematic of this situation, which also illustrates the expected importance of confining boundaries.

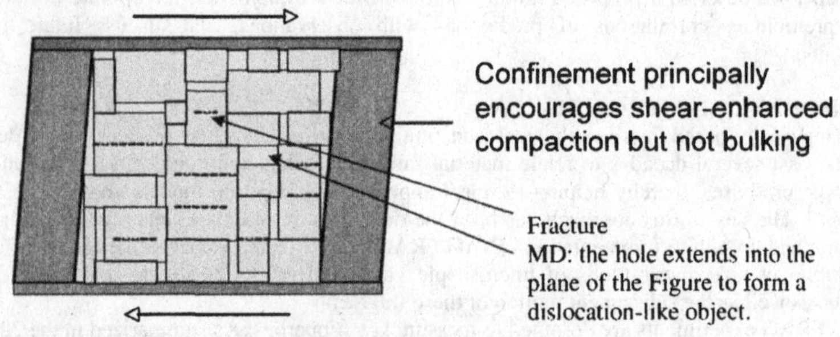

Figure 1. Schematic picture of conceptual mesomodel

The proposed unpinning criterion is based on a simplification of the FB2 comminution model. We assume that an applied remote "driving stress" state (τ, P) is sufficient to fracture fragments of size B_u or greater, given by

$$\eta B_{II} = \pi K_{Ic}^2/4(m_1\tau^2 + m_2P^2) \tag{1}$$

where τ is the maximum shear stress, P is the pressure, B_{II} is a critical fragment size, ηB_{II} is the corresponding critical flaw size, K_{Ic} is the plane strain fracture toughness, and the m's allow local stress enhancement over the remote stresses. Figure 1 shows that we idealize the fragment cross-sections as squares. An equivalent circle with the same area would thus have a radius $R = B/\pi^{1/2} = 0.56B$.

Eq (1) is clearly an oversimplified relation to be improved as more data are obtained. For example, the dynamic initiation or arrest toughness and/or a combination of Mode I and Mode II toughnesses would be more appropriate than K_{Ic}, which is used here as a simple measure of brittleness, and to show trends. We expect that a more detailed model, such as that of Simons et al[9] for concrete and marble, may eventually be needed.

As the applied stresses increase, comminution breaks down the larger fragments until either sufficient unpinning has occurred to allow flow of the remaining fragments, or comminution ceases because the flaw sizes are subcritical.

Specifically, we fit our data to an initial Poisson fragment size distribution

$$N_g(B) = N\exp(-B/B_0) \tag{2}$$

where $N_g(B)$ is the number of fragments per unit area of a cross section with size greater than B, N is the total number of fragments per unit area in a cross section, and B_0 is the initial average size of the fragments[4]. The fragment density function is

$$dN/dB = (N/B_0)\exp(-B/B_0) \tag{3}$$

We assume that the initial distribution has an upper cutoff, a largest fragment, B_{max}. We also assume that the initial hole (MD) size distribution mirrors the fragment size distribution, with the larger holes being associated with the larger fragments, and the average hole size is equal to the average fragment size. Integrating the hole area bB^2 with a density function like Eq(3) from B = 0 to B_{max} gives the total initial porosity

$$\phi_T = N_H bB_0^2[2 - f(x)] \tag{4a}$$

where N_H is the total number of holes per unit area of a cross section, b is the fraction of B that specifies the width of the hole, i.e. bB is the macroscopic Burger's vector, and

$$f(x) = (x^2+2x+2)\exp(-x) \tag{4b}$$

where $x = B_{max}/B_0$.

For a given B, say B_1, the mobile porosity is the total porosity minus the integral from 0 to B_1, and the ratio of mobile to total porosity is

$$\phi_M/\phi_T = [f(x_1) - f(x_2)]/[2 - f(x_2)] \tag{5}$$

where $x_1 = B_1/B_0$, and $x_2 = B_{max}/B_0$.

For example, Eq(5) shows that when $B_1 = B_0$ and $B_{max} = 2\,B_0$, $\phi_M/\phi_T = 0.75$. That is, when the largest fragment in the distribution has been reduced to B_0, 75% of the original fragments and associated MDs have been unpinned.

For such flow to be possible, the material must contain pores (the MDs). Under the high confinement provided by the impact interface of a uniaxial strain plate impact, or on-axis at the nose of a rod impact, the MDs will tend to be driven into the confining boundary, resulting in compaction (the fragments can move into the RVE, but not out). If the confinement is maintained, the subsequent response must be elastic. This postulated behavior is shown schematically in Figure 2.

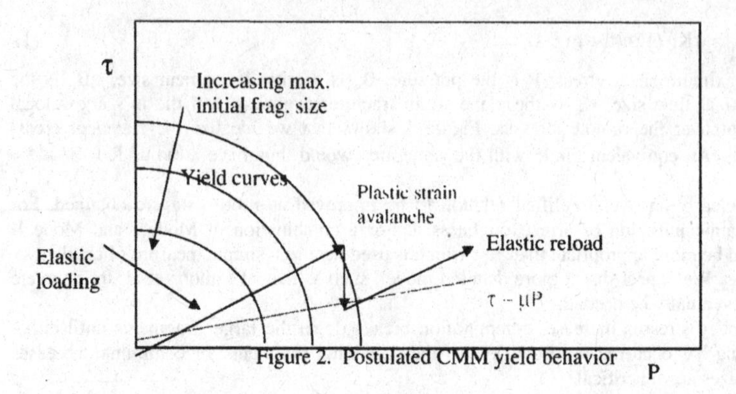

Figure 2. Postulated CMM yield behavior

The mesomodel must also describe the process by which tensile cracks and MDs (voids) are produced in originally void-free glass under compression and shear. "Wing cracks" are a candidate for causing brittle failure and dilatancy under compression and shear (see Figure 3). In general, a weak shear surface flaw can either propagate as a Mode II shear crack, or turn out of the crack plane and propagate as a wing ("splitting") crack. In the latter case, voids (MDs) are produced in the target, causing dilatancy. The extensive literature on this subject is reviewed in the 2004 book by Kanel, Razorenov, and Fortov[5]. Wing cracks were observed in glass plates under compression in 1963 by Brace and Bombolakis[10]. Subsequent work by Nemat-Nasser and Horii[11], Horii and Nemat-Nasser[12], Moss and Gupta[13], and Nemat-Nasser and Obata[14], among others, described expected behavior under different stress states. Kalthoff[15] performed experiments with an edge impact technique on a number of ductile and brittle materials, and found that the mode chosen depended on whether there was a mechanism for shear softening (e.g. adiabatic heating) sufficient to stabilize a propagating shear crack.

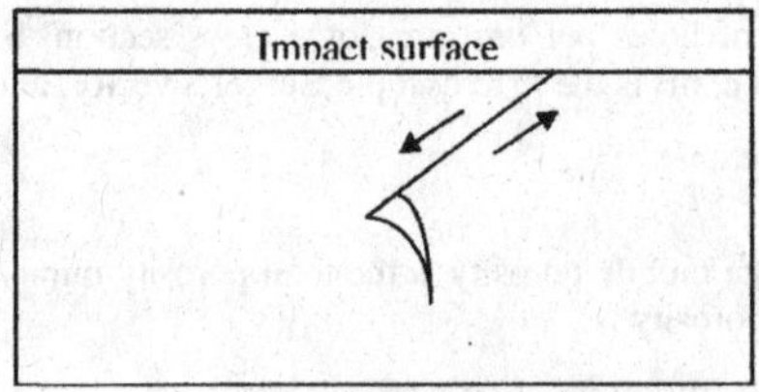

Figure 3. Wing crack

Thus, wing cracks appear to be a possible source for driving voids (MDs) into previously non-porous glass in plate impact tests. However, the brittle-ductile transition discussed above may suppress their formation, in which case we would expect a non-porous layer of damage at the impactor-target interface. To resolve this and other issues, we need more data.

SUPPORTING EXPERIMENTS

Uniaxial strain plate impact response:

We begin by examining the response of Soda-Lime Glass (SLG) samples loaded by uniaxial strain impacts in experiments reported by Simha and Gupta[16], and Alexander et al[17]. We assume that cracks (wing cracks or shear cracks), originate from a size distribution of flaws of size δ on the impacted surface. We also assume that the distribution is of the form of Eq(2). To relate the flaw size to the fragment size B, we draw on the "crack range" concept of the BFRACT model for tensile cracks[4], which defines a parameter $M = B/\delta$. For very brittle materials, M can be between 10 and 20. The FB2 comminution model assumes that each fragment contains a flaw of size $\delta = \eta B$. Thus, $M = 1/\eta$.

For each active slip plane, Eq (1), with δ set equal to the minimum flaw size, describes the Fig. 2 yield function in $\tau - P$ space. When an elastic load path crosses the surface, the model produces a burst of non-elastic strain as the porosity is driven into the impactor surface. The shear stress drops at constant pressure to the "failed" curve $\tau = \mu P$.

To describe the non-elastic flow of the "yielded" material in a manner that ensures stability and uniqueness, an analysis due to Whitham[18,19] is applied, and a "stress-relaxing solid" relation is used to describe the total strain rate on a slip plane as the sum of the elastic strain rate and the non-elastic strain rate:

$$\partial\varepsilon/\partial t = (1/2G)\,\partial\tau/\partial t + \partial\varepsilon^{ne}/\partial t \qquad (6a)$$

where the non-elastic strain rate is given by

$$\partial\varepsilon^{ne}/\partial t = [\phi/2(2b)^{1/2}B][(\tau-\mu P)/\rho]^{1/2}\,H(\tau-\mu P) \qquad (6b)$$

and where G is the shear modulus, H is the Heaviside function, τ is the maximum shear stress ($=\sigma_{eq}/2$), μ is the intergranular friction coefficient, ϕ is the mobile (unpinned) porosity associated with the MDs, B is the average hole (MD) height (equal to the average fragment size), and b is a dimensionless parameter that characterizes the average MD width. The quantity $[\phi/(b)^{1/2}B]$ can be considered a meso-parameter that specifies, for a given stress state on a given slip plane, both the porosity carried by a MD and the MD speed. It has significant leverage on the predicted behavior, and we will vary it in a parameter study.

In Table 1, we list measured and assumed material properties for the Soda-Lime Glass (SLG), and compare them with those for B_4C, a ceramic for which a clear two-wave response has been measured[20]. The glass and B_4C data warrant continued study, since both sets of experiments were well-instrumented, and give an opportunity to study the effects of different microstructures, fracture toughnesses, and moduli.

Table 1. Properties for SLG and B4C

PROPERTY	SLG	B4C
Density (g/cc)	2.5	2.51
C_l (km/s)	5.761	13.7
C_s (km/s)	3.437	8.7
C_{BULK} (km/s)	4.176	9.3
ν (Poissons ratio)	0.224	0.162
G	12 MPa* (29.5 GPa)	190 GPa
K	17 MPa* (43 Gpa)	218 GPa
E	29 MPa* (73 GPa)	
K_{Ic} (MPa-m$^{1/2}$)	0.9	4
HEL (GPa)	3.5 – 7	16.15
Flaw locations	Surface	Surface and internal; grain boundaries, inclusions, etc.
Assumed meso properties, μ, η, B_l, B_{0}, B_{max}, $\phi/b^{1/2}B_0$	0.3, 0.1, 1μm, 50 μm, 100μm, 10 - 100 cm^{-1}	

* The reported[17] ambient pressure elastic moduli measured by ultrasound are inconsistent with the measured wave speeds, and have been corrected. The corrected values are given in the parenthesis.

We do not have measurements of the surface flaw distributions in the glass, but we will do an example analysis assuming the mesomechanical properties listed in Table 1, including setting B_1 = 1 μm, B_0 = 50 μm, and B_{max} = 100μm. We next apply the simplified flow model of Eq (6b).
To convert the τ relations to those for the longitudinal stress, S, we use the uniaxial strain conditions:

$$\tau = [3(1\text{-}2\nu)/2(1+\nu)] \quad P = P/\psi, \text{ where } \psi = [2(1+\nu)/(3(1-2\nu)], \text{ and } S = [2(1\text{-}\nu)/(1-2\nu]\tau \tag{7}$$

Operating on Eq (6a) with ∂/∂t, combining with the equations for conservation of mass and momentum, and using the above uniaxial strain relations connecting S and τ via Poisson's ratio ν, leads to

$$\partial^2 S/\partial t^2 - C^2 \partial^2 S/\partial h^2 + \lambda(\partial S/\partial t) = 0 \tag{8}$$

where h is the Lagrangian distance into the target, C is the longitudinal wave speed,

$$\lambda = (1/T) = [(1\text{-}2\nu)/2(1-\nu)]^{1/2}\,[1-\mu\psi]^{1/2}\,[0.18\phi/b^{1/2}B]\rho^{1/2}C^2S^{-1/2} \tag{9}$$
and $\psi = [2(1+\nu)/(3(1-2\nu)]$.

Using the glass values of Table 1 yields values of T = ranging from 50 to 500 ns, depending on the choice of $\phi/b^{1/2}B_0$.

Eqs (8) and (9) fulfill a stability criterion due to Whitham[18,19]: If λ • 0, the solution to Eq(8) for S(h,t) is well-posed and stabile (dissipative). The result is a decaying elastic wave followed by a diffusive failure wave. In the linear approximation, S is given by

$$S \bullet S_0(t\text{-}h/C)\exp[-\lambda h/2C] \tag{10}$$

where S_0 is the impact stress. Thus, the width of the MZ is approximately 2CT. λ in Eq(9) is not a constant, but depends on the stress to the -1/2 power, thereby violating the linear assumption. However, the "viscous" overshoot from the HEL is small, so we will roughly approximate the response by setting S equal to the HEL value.

We now focus on specific experiments with SLG glass performed by Simha and Gupta and Alexander et al, in which SLG impactors on SLG targets generated elastic impact stresses of 4 to 10 GPa at the impact surface. Both measured loading times to 4 GPa of about 0.2μs. For impact stresses of 4 to 6 GPa, Simha and Gupta measured a two-wave structure. For example, at 4.6 GPa, they measured a slightly rounded longitudinal stress plateau at 4 GPa, followed after about 2 μs by a second rise to 4.6 GPa. This can be combined with their lateral stress history record to show a strength, τ, that jumps to a plateau of about 1.5 GPa, but after about 0.5 μs, drops to about 1 GPa, only to ramp up again at about 2.5 μs to a new plateau of about 1.5 GPa at about 3 μs. This observed two-wave structure was interpreted by Simha and Gupta as a time-dependent loss of strength followed by a partial regaining of that strength, consistent with the CMM picture. In contrast, Alexander et al did not record a two-wave structure.

Simha and Gupta developed an ad hoc continuum strength model that correlated well with the above experiments, has similar features as the CMM model, and can therefore serve as a test of the CMM model's validity.

That is, to compare the CMM predicted trends with the above data, we set $m_1 = m_2 = 1$, and choose the parameters in Table 1 to enforce a value of 4 GPa for the HEL. The values in Table 1 for SLG give a uniaxial strain loading path of τ = 0.68P, and Eq (1) gives a yield circle with a radius = 2.5

GPa, corresponding to S (HEL) = 4 GPa, as desired. The CMM model calculates a diffusion front pseudovelocity obtained by analogy to heat flow calculations[4], where calculations of heat flow from a hot slab, maintained at constant temperature, suddenly placed in contact with a cold material, showed that the pseudovelocity of propagation of one-fourth the hot slab temperature, was 3k/h, where k is the diffusivity. Taking the diffusivity to be equal to C^2 T, and setting the pseudovelocity equal to h(diffusion)/t yields

$$h(\text{diffusion}) = 1.7C(Tt)^{1/2} \quad (11)$$

So far, the discussion has concerned individual slip planes, but a further consequence of the postulated "unpinning yield condition" is that many slip planes would become active simultaneously. In Figure 1, for example, the vertical slip planes would also start to slide. The fragment cross sections, schematically shown as squares, would become "rounder", and the material would become more like a liquid. The effective coefficient of friction μ might decrease as the particles begin to roll. But if the confinement is maintained, once the MDs have flowed into the confining penetrator interface to compact the material, the subsequent reloading from the relaxed state S ($\tau = \mu P$) of 1.74 GPa (for μ = 0.3) would be that of an elastic liquid, and the appropriate wave speed would be the bulk wave speed.

Since the CMM model has many adjustable parameters, a wide variety of responses can be predicted. We performed a preliminary parameter study by varying the value of $\phi/b^{1/2}B_0$ over the range shown in Table 1. Figure 4 shows the two extremes (T = 50 ns and T = 500 ns) for the Simha and Gupta experiment which produced an impact stress of 4.6 GPa. The best overall correlation is obtained with the large value of T (500 ns), which forced the diffusing failure wave to travel at almost the elastic wave speed. The slower bulk wave speed for the elastic liquid delayed the arrival of the reloading wave to about 1.6 μs, in rough agreement with the Simra and Gupta results for the longitudinal stress. However, the observed regaining of strength is not predicted by the CMM (although it is possible to imagine that the fine particles compact and "freeze" to become an effective solid again).

By choosing the small value of T (50 ns), we can delay the reloading pulse to agree with the second wave arrival time recorded by Simha and Gupta. However, Figure 4 shows that the delay simply reduces the first wave to low amplitude, resulting in a poor correlation.

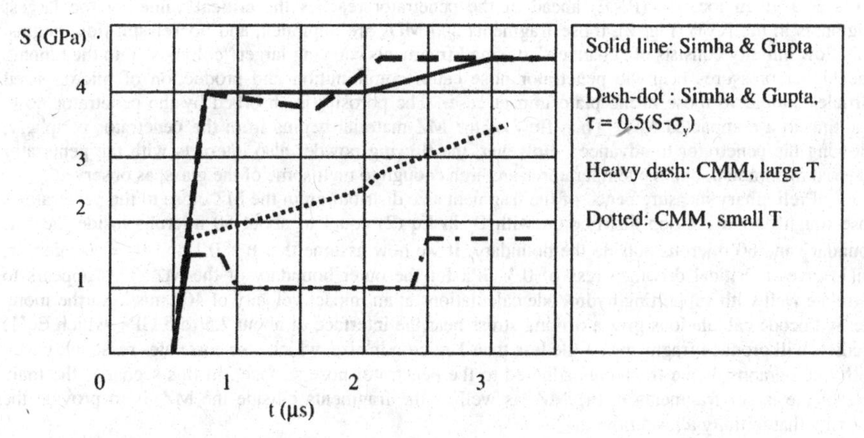

Figure 4. Comparison of CMM model trends with simplified Simha & Gupta data. Measurement is 3 mm from impact surface.

At higher impact stresses, the data suggest that the glass becomes ductile, and/or undergoes densification or a phase change, and no longer behaves like a collection of elastic fragments. Since our model predicts that the material will recompact at the impactor interface, and reload elastically, we expect that both reshocking and unloading should occur elastically, as diagrammed schematically in Figure 2. Elastic unloading was, in fact, inferred by Alexander et al[17].

In summary, there are enough adjustable parameters in the CMM model to allow rough trend correlations with the plate impact data to be obtained, but we need additional damage evolution data to further constrain our model parameters. A proposed "soft recovery" plate impact experiment will be described later.

Partial penetration experiments provide valuable data under other loading conditions, and are discussed next.

Partial penetration of non-eroding rods:

As described in a companion paper in the present conference[2], we have performed experiments in which hard steel hemispherical-nosed rods were fired into thick, confined soda lime and borosilicate glass targets at impact velocities ranging from 300 to 600 m/s. The 6.35 mm diameter x 31.8 mm long rods remained elastic, and arrested after partial penetration. The recovered targets showed a MZ region around the penetrator that consisted of pulverized material with a fairly sharp boundary. The MZ thickness at the nose at arrest in all cases was less than 2 mm. The diameter of the tunnel ranged from about twice the rod diameter at 300 m/s impact velocity to about 4 times the rod diameter at 600 m/s impact velocity.

The membrane stresses nucleate finely-spaced cone cracks which are driven into the target from the rod periphery as the rod advances, and intersect with less finely-spaced lateral cracks to form a fragmented bed ranging from mm-sized fragments some distance ahead of the arrested penetrator to sub mm-sized fragments close to the boundary with the MZ, within which the fragment sizes were less than 50 microns. The cone cracks open to relieve the stress and produce porosity (MDs). Thus, similar to the potential role played by wing cracks in the plate impact case, a mechanism exists to produce both fragments and porosity (MDs) in a region ahead of an oncoming interface with the impenetrable penetrator or flyer plate.

The CMM model then suggests the following penetration scenario. When the axial compressive stress at a given location (RVE) ahead of the penetrator reaches the critical value for the largest fragments in the RVE (Fig. 2), those fragments and MDs are unpinned, and non-elastic flow begins. This flow initially consists of "channels" of small fragments carrying larger "colonies" into the tunnel[2]. The higher pressures near the penetrator nose cause comminution and production of micron-sized particles similar to those in the plate impact case. The porosity is absorbed by the penetrator nose, resulting in a compacted MZ. Then flow of the MZ material begins from the penetrator periphery, allowing the penetrator to advance. However, the flowing powder also interacts with the penetrator material to abrade it, and the temperatures are high enough to melt some of the glass, as observed[2].

Preliminary measurements[2] of the fragment size distribution in the MZ close to the penetrator's nose roughly fit a Poisson distribution with B_o in Eq (2) equal to about 50 microns inside the MZ boundary and 60 microns outside the boundary. If we now assume that η is 0.1 and B_o = 50 microns, Eq(1) gives a critical driving stress of 0.3 GPa for the outer boundary of the MZ. This appears to correlate well with supporting hydrocode calculations at an impact velocity of 400 m/s[21]. Furthermore, the hydrocode calculations give a driving stress near the interface of about 2.5 to 3 GPa, which Eq(1) predicts will produce fragments a little less than 1 micron in size, which also correlates reasonably well with micrographs of the fragments adhered to the penetrator nose-surface[2]. In this scenario, the main role of the larger fragments in the MZ, as well as the fragments outside the MZ, is to provide the porosity that initially allows flow.

Thus, the CMM correlations are perhaps encouraging, but we again need more microdamage evolution data, as discussed next.

FUTURE TEST PROGRAM

Partial penetration tests:

The partial penetration tests fulfill the basic requirements for measuring microdamage evolution: variable load amplitudes and durations, soft-recovered specimens allowing microscopic examination of the fractured material, and recovered tunnel material for property testing. A prototype test program is described in a companion paper in this conference[2].

Quasistatic property tests:

Tests are needed to obtain basic material properties, especially for the pulverized material, for input to both continuum and mesomechanical models. Pressure–shear measurements on pulverized material recovered from the partial penetration tests are underway[2]. Those tests include microstructural observations of the material before and after granular flow. We plan similar observations of damaged material from confined pressure tests being performed at SwRI. Since interaction of the pulverized material debris with the penetrator nose and sides seemed important in the partial penetration tests, we plan to also examine the material that adheres to the shear surfaces in the SwRI specimens. We will compare the evolution of fragment size and shape for the two cases.

Soft-recovered plate impact tests:

To help interpret prior plate impact data, we need tests that allow us to measure the evolution of the microdamage. In prior work on brittle tensile fracture and fragmentation in Armco iron, for example, we were able to produce different damage levels in target "pucks", and thereafter perform iterative calculations with NAG/FRAG models until we could correlate with the measured damage distributions[4]. To follow the same procedure for compression-shear loads, we need a scheme to soft-recover the target specimens for subsequent microscopic examination. A possible design is sketched in Figure 5, which shows the following features based on our earlier work[3,4]:

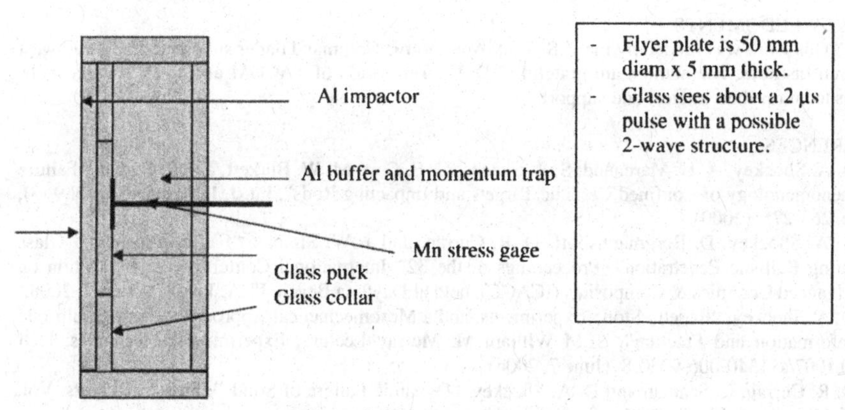

Figure 5. Plate impact test design

- The samples are intact or pre-fractured cylindrical glass plates ("pucks").
- Impactor and confinement materials are impedance-matching aluminum alloys.
- The sample plate is surrounded by a glass collar whose purpose is to eliminate converging unloading waves, and maintain uniaxial strain in the sample for the duration of the load pulse.
- Impact velocities and transmitted stress histories are measured.
- Increasing impact velocity levels produce increasing degrees of damage.
- A standard "rag cage" is used for soft-recovery of the target package.
- The glass sample is characterized pre and post-test for fragment size, pore size, and fragment geometry distributions.

CONCLUSIONS

Our conceptual mesomodel (CMM) shows some promise, but needs more microdamage evolution data to improve it. As discussed above, we expect that Eq(1) will be replaced by an expression that contains static and dynamic Mode I and Mode II toughnesses and other refinements, and sophisticated crack growth and coalescence models are available for correlating better with microdamage evolution data. Some key questions are:

- How is dilatancy introduced in impacted glass targets free of internal flaws? Are wing cracks the actual mechanism? The proposed experiments of Figure 5 should help us answer that and other questions.
- What governs the sudden reduction of fragment sizes inside the MZ? The picture of "strain burst"–induced compaction followed by elastic reloading under confinement and compaction will no doubt be modified by data from the proposed additional partial penetration tests and plate impact tests.
- What are the flow properties of the material in the MZ? Data from the planned quasistatic property tests will be valuable in this area.

In general, the forthcoming detailed microscopic damage evolution data should help us replace uncertain elements of our conceptual mesomodel with empirically-determined relations.

ACKNOWLEDGMENTS

This work was funded by the U.S. Tank Automotive Command under subcontract to Southwest Research Institute. The authors are grateful to D. W. Templeton of TACOM and C. E. Anderson, Jr. and his team for their interest and support.

REFERENCES

[1]D. A. Shockey, A. H. Marchand, S. R. Skaggs, G. E. Cort, M. W. Burkett, and R. Parker, "Failure Phenomenology of Confined Ceramic Targets and Impacting Rods", Int. J. Impact Engr, Vol 9 (3), pp. 263-275, (2000).

[2]D. A. Shockey, D. Bergmannshoff, D. R. Curran, and J. W. Simons, "Failure Physics of Glass during Ballistic Penetration", Proceedings of the 32nd International Conference & Exposition on Advanced Ceramics & Composites (ICACC), held at Daytona Beach, FLA, Jan. 27 – Feb. 1, 2008.

[3]D. A. Shockey, "Rosetta Stone Experiments, and a Mesomechanical Approach to High Strain-rate Deformation and Fracture", SEM William M. Murray Lecture, Experimental Mechanics, DOI 10.1007/s11340-006-9030-8, (June 7, 2006).

[4]D. R. Curran, L. Seaman and D. A. Shockey, "Dynamic Failure of Solids", Physics Reports, Vol. 47, Nos. 5 & 6, March 1987.

[5]G. I. Kanel, S. V. Razorenov, and V. E. Fortov, Shock-Wave Phenomena and the Properties of Condensed Matter, Springer-Verlag, New York (2004).

[6]Th. Behner, Ch. E. Anderson Jr., D. L. Orphal, M. Wickert, V. Hohler, and D. W. Templeton, "Failure and Penetration Response of Borosilicate Glass during Short Rod Impact", Proc. 23rd Int. Symp. Ballistics, **2**, pp 1251-1258, Graficas Couche, Madrid, Spain, (2007).
[7]C. E. Anderson Jr., T. Behner, T. J. Holmquist, M. Wickert, V. Hohler, and D. W. Templeton, "Interface defeat of long rods impacting borosilicate glass", Proc. 23rd Int. Symp. Ballistics, **2**, pp 1049-1056, Graficas Couche, Madrid, Spain, (2007).
[8]D. R. Curran, "Comparison of Mesomechanical and Continuum Granular Flow Models for Ceramics", Proceedings of the APS SCCM Topical Conference, 31 July-6 August 2005, Baltimore Md.
[9]J. W. Simons, T. H. Antoun, and D. R. Curran, "A Finite Element Model for Analyzing the Dynamic Cracking Response of Concrete", Presented at the 8th International Symposium on Interaction of the Effects of Munitions with Structures, McClean, Virginia, (April 22-25, 1997).
[10]W. F. Brace and E. G. Bombolakis, "A Note on Brittle Crack Growth in Compression", J. Geophys. Res. **68**, pp. 3709-3713, (1963).
[11]S. Nemat-Nasser and H. Horii, "Compression-induced Nonplanar Crack Extension with Application to Splitting, Exfolation, and Rockburst", J. Geophys. Res. **87(B8)**, pp. 6805-6821, (1982).
[12]H. Horii and S. Nemat-Nasser, "Compression-induced Macrocrack Crack Growth in Brittle Solids: Axial Splitting and Shear Failure", J. Geophys. Res. **90(B4)**, pp. 3105-3125, (1985).
[13]W. C. Moss and Y. Gupta, "A Constitutive Model Describing Dilatancy and Cracking in Brittle Materials", J. Geophys. Res. **87(B4)**, pp. 2985-2998, (1982).
[14]S. Nemat-Nasser and M. Obate, "A Microcrack Model of Dilatancy in Brittle Materials", Trans. ASME: J. Appl. Mech. **55(110)**, pp. 24-35, (1988).
[15]J.F. Kalthoff, "Modes of dynamic shear failure in solids", Int. J. Fracture **101**, pp. 1-31, (2000).
[16]C. H. M. Simha and Y. M. Gupta, "Time Dependent Inelastic Deformation of Shocked Soda-lime Glass", J. Appl. Physics, Vol. 96, No. 4, pp 1880-1890, (15 August, 2004)
[17]C. S. Alexander, L. C. Chhabildas, and D. W. Templeton, "The Hugoniot Elastic Limit of Soda-lime Glass", in Shock Compression of Condensed Matter – 2007 (M. Elert, M. D. Furnish, R. Chau, N. Holmes and J. Nguyen eds., AIP Press, pp 733-738 (2007).
[18]G. B. Whitham, " Some Comments on Wave Propagation and Shock Wave Structure with Application to Magnetohydrodynamics", Comm. Pure and Applied Mathematics, Vol. XII , 113-158, (1959)].
[19]G. B. Whitham, Linear and Nonlinear Waves, John Wiley & Sons, 1974.
[20]T. J. Holmquist and G. R. Johnson, "Characterization and Evaluation of Boron Carbide for 1D Plate Impact", Tech. Report: SwRI report 18.10174/03, Contract F42620-00-D-0037-BR02, (July 2006).
[21] C. E. Anderson, personal communication.

OPTIMIZING TRANSPARENT ARMOR DESIGN SUBJECT TO PROJECTILE IMPACT CONDITIONS

Xin Sun* and Kevin C. Lai
Pacific Northwest National Laboratory
K6-08
PO BOX 999
Richland, WA 99352

Tara Gorsich and Douglas W. Templeton
USARMY RDECOM-TARDEC
Warren, Michigan 48397

ABSTRACT

Design and manufacturing of transparent armor have been historically carried out using experimental approaches. In this study, we use advanced computational modeling tools to perform virtual design evaluations of transparent armor systems under different projectile impact conditions. AHPCRC developed modeling software EPIC'06 [1] is used in predicting the penetration resistance of transparent armor systems. LaGrangian-based finite element analyses combined with particle dynamics are used to simulate the damage initiation and propagation process for the armor system under impact conditions. It is found that a 1-parameter single state model can be used to predict the impact penetration depth with relatively good accuracy, suggesting that the finely comminuted glass particles follow the behavior similar to a viscous fluid. Even though the intact strength of borosilicate and soda lime glass are different, the same fractured strength can be used for both glasses to capture the penetration depth.

INTRODUCTION

The response of brittle solids to the impact loading of different projectiles has been a subject of considerable interest because of its significance to strength degradation and structural integrity of this important class of materials. It becomes particularly important in designing transparent armor materials because of the seemingly conflict goals of providing ballistic protection and maintaining vehicle weight and mobility. Design and manufacturing of transparent armor have been historically carried out using experimental approaches. This is because glass strength under tensile loading condition is controlled by surface flaws and therefore it is difficult to assess glass structural integrity using a deterministic analytical approach.

In order to examine penetration resistance of various glasses under ballistic impact condition, many researchers have conducted controlled impact experiments. For example, Behner et al. [2] measured failure front dynamics in borosilicate glass under gold rod impact and penetration. The dynamic failure wave front and rod penetration depth are measured simultaneously using high speed photography and flash X-rays. Nie et al. [3] measured the dynamic strength of borosilicate glass under combined compression and shear loading using Split-Hopkinson Pressure Bar (SHPB) technique. More recently, Bless and Chen [4] reported the experimental ballistic data for layered soda lime glass under the impact loading of a fragmentation simulated projectile (FSP). Even though penetration is limited in the top two glass layers, various damage patterns extends throughout every plate. With more experimental data becoming available, modeling activities are focused on finding the appropriate material constitutive model and the associated material parameters in modeling and predicting the impact resistance of glass. For example, Sun et al. [5] used a continuum damage mechanics (CDM)-based constitutive model to describe the initial failure and subsequent stiffness reduction of glass of the experiments carried out by Nie et al. [3]. A maximum shear stress-based damage evolution law was used in describing the glass damage process under combined compression/shear loading. With only two modeling parameters, reasonably good comparisons between the predicted and the experimentally measured failure maps have been obtained for various glass sample geometries under the impact strain rate of 250/sec. Under higher strain rate and higher confinement pressure, Johnson and Holmquist [6] examined the applicability of various constitutive models in EPIC'06 for the gold rod penetration experiments carried out by Behner et al. [2]. It was found that a 1-parameter single state model can be used to predict the penetration velocity as well as the hole profile and penetration depth with reasonably good accuracy.

In this study, we examine the applicability of the 1-paremeter single state model in EPIC'06 [1] in modeling the FSP penetration experiment reported by Bless and Chen [4] on layered soda lime transparent armor systems. Both monolithic glass and layered glass armor are examined with EPIC'06. Observations are then made on the predicted penetration depth and damaged glass strength for borosilicate and soda lime glass with different constitutive models. Directions for future research are also suggested.

GLASS CONSTITUTIVE MODEL VALIDATION WITH FSP SINGLE SHOT EXPERIMENT

AHPCRC developed modeling software EPIC'06 [1] is used in this section to study the applicability of a simple 1-parameter single-state constitutive model to model the FSP penetration experiment reported by Bless and Chen [4]. LaGrangian-based finite element analyses combined with particle dynamics are used to simulate the damage initiation and propagation process for the soda lime glass under different impact conditions. The fractured material strength is modeled with only 1 parameter: constant equivalent stress of 0.570GPa, see the horizontal line illustrated in Figure 1.

At the first glance, this model may seem over simplified since it assumes that the fractured glass strength does not depend on confinement pressure or impact strain rate. From another perspective, this assumption may be physically sound since it describes an elastic-perfectly plastic behavior, i.e., viscous flow condition provided by the finely comminuted glass particles immediately in front of the projectile [7]. It should be mentioned that various other constitutive models have also been examined by Johnson and Holmquist [6] to simulate the gold rod penetration test on borosilicate [2], and that the simple 1-parameter model was found to be able to predict the penetration velocity and depth with reasonably good accuracy. The value 570MPa is obtained by matching the gold rod penetration test data at V = 1600 m/s, see Figure 2 [2, 6].

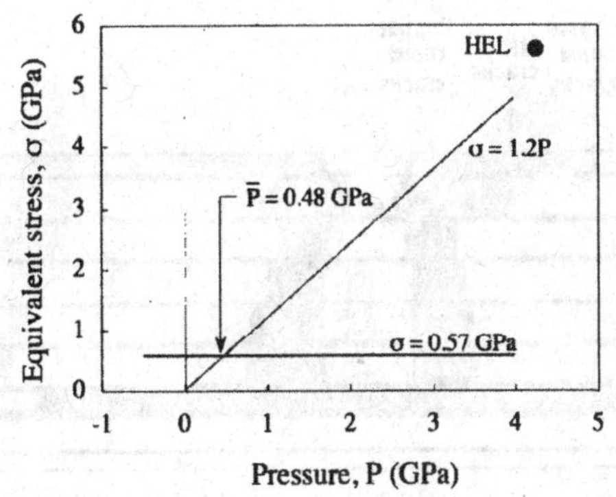

Figure 1. Single-state model for strength of fractured glass (re-plotted from Johnson and Holmquist [6])

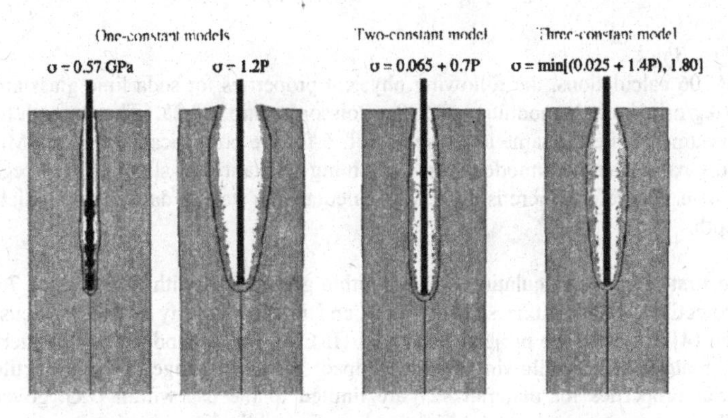

Figure 2. Predicted hole profiles of the single state model at t = 28 μs for V = 1600 m/s (re-plotted from Johnson and Holmquist [6])

The single-state 1-parameter glass model is next used in modeling the FSP penetration experiment reported by Bless and Chen [4] on soda lime glass. The target used in the experiment was made of seven layers of 30cm x 30cm soda lime glass of 7.6cm total thickness and two layers of polycarbonate of 12.7mm total thickness. The individual layers were assembled with polyurethane adhesive. A standard 12.7mm (50-caliber) fragment simulating projectile was shot into the glass layers at 1118m/s. The projectiles weigh 13.4g and are made of HRC30 steel [4]. The projectile was stopped in the second glass layer, with the total depth of penetration of 18.9mm. Figure 3 is a replot from Ref. 4 showing the different crack morphologies observed for the entire target.

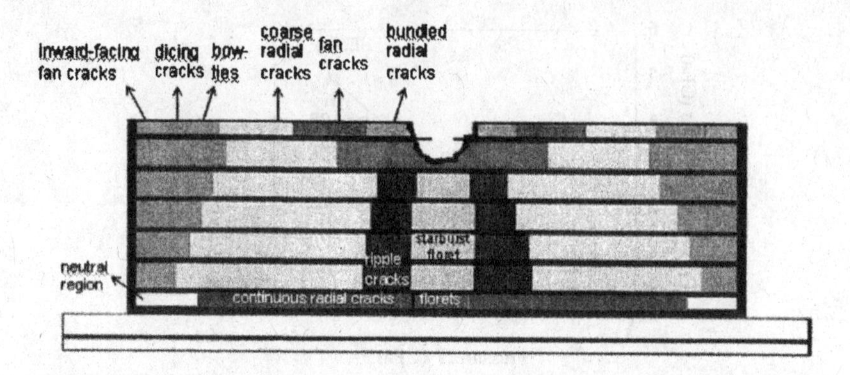

Figure 3. Damage map of complete target for the single shot FSP experiment (re-plotted from Ref. 4)

Monolithic Glass Block

In EPIC'06 calculations, the following physical properties for soda lime glass are used: density = 2530kg/m^3, Young's modulus = 72GPa, Poisson's ratio = 0.25. The strength for soda lime glass is assumed to be the same as those in Ref. 6 for the borosilicate glass: 570MPa, and the single-state 1-parameter glass model is used assuming constant flow stress of glass regardless of its damage state. Therefore, there is no need to calculate the state of damage in predicting the penetration depth.

For the first round of calculations, a monolithic glass target with thickness of 76mm is used. The projectile with the same shape, weight and impact velocity as those discussed by Bless and Chen [4] is used. The projectile material HRC30 steel is modeled with material #27 from the EPIC material library allowing possible impact induced damage to the projectile. The detailed material properties for material #27 are limited to the use within U.G. government agencies and their contractors and can not be listed here for public dissemination.

Figure 4(a) shows the predicted impact region profile and damage contour for the projectile. The predicted penetration depth is 18.39mm, slightly lower than the experimentally observed value of 18.9mm for the bonded glass layers. Some damage is predicted for the projectile, but the damage values are less than 1 (red in contour). Again, since no damage is calculated with the simple 1-parameter model, the predicted penetration plot in Figure 4(a) does not show any damage on it.

For comparison purpose, the brittle material model #82 in EPIC'06 library for float glass is also used in modeling the monolithic glass target performance. Material #82 is a typical brittle material modeled with JH-2 model [1]: the strength for the JH-2 model is expressed in a normalized analytic form, and that the strength is degraded gradually as the damage accumulates. The intact strength at D=0 is expressed as:

$$\sigma_i^* = A(P^* + T^*)^N (1 + C \ln \dot{\varepsilon}^*) \qquad (1)$$

And the fractured strength at D=1.0 is:

$$\sigma_f^* = B(P^*)^M (1 + C \ln \dot{\varepsilon}^*) \quad (2)$$

The general strength expression for damaged material is:

$$\sigma^* = \sigma_i^* - D(\sigma_i^* - \sigma_f^*) \quad (3)$$

The dimensionless strength quantities are normalized by the equivalent stress at the HEL, in the general form $\sigma^* = \sigma / \sigma_{HEL}$. In EPIC'06, the default material constants for material #82 are: A=0.93, B=0.088, C=0.003, M=0.35 and N=0.77. The predicted damage zone is shown in Figure 4(b) in red color, and the penetration depth is 24.10mm, much higher than that of the 1-paramter model prediction.

Layered Glass Armor

Next, the effects of the polyurethane adhesive layer on the impact resistance of layered armor are examined. The layering scheme for the laminated glass is the same as those used by Bless and Chen [4] with seven layers, and the thickness of each glass layer from top to bottom is: 6.8mm, 12.48mm, 12.48mm, 12.48mm, 12.48mm, 12.48mm and 6.8mm. Two polycarbonate liners are also used at the bottom of the target, and their respective thickness is 8mm and 4.7mm. The thickness of the PVB interlayer is constrained by the minimum thickness requirement in EPIC: 2.12mm.

Again, both the 1-parameter constant flow stress model and the library material model #82 are used in modeling the soda lime glass. The polyurethane adhesive is modeled with EPIC library material #142 for polyvinyl butyral (PVB) phenolic resin [1] with the Johnson-Cook model [11]:

$$\sigma = [C1 + C2 \cdot \varepsilon^N][1 + c3 \cdot \ln \dot{\varepsilon}^*][1 - T^{*M}] + C4 \cdot P \quad (4)$$

where ε is the equivalent plastic strain, $\dot{\varepsilon}^*$ is the dimensionless total strain rate, T* is the homologous temperature, and P is the hydrostatic pressure (compression is positive). The default constants for material #142 in EPIC'06 are: C1= 75.8MPa, C2=69MPa, N=1.0, C3=0, M=1.85 and C4=0.

Figure 5 shows the predicted impact zone for the two glass models with the PVB interlayers. The predicted penetration depths are: 18.97mm for the 1-parameter model, and 27.38mm for the library material #82. For both glass models, the predicted total penetration depths for the layered armor are higher than their monolithic counterparts. The results in Figures 4 and 5 also indicate that the predicted penetration depth with the simple 1-parameter model is very close to the experimentally measured penetration depth of 18.9mm for layered soda lime glass. Table 1 compares the depth of penetration (DOP) measured in the experiment with DOP predicted by the 1-parameter model and by material model #82.

Table 1. Comparisons between predicted and measured DOP

DOP in mm	Experimentally measured	EPIC'06 prediction 1-parameter model	EPIC'06 prediction material #82
Laminated glass	18.9	18.97	27.38
Monolithic glass	Not performed	18.39	24.10

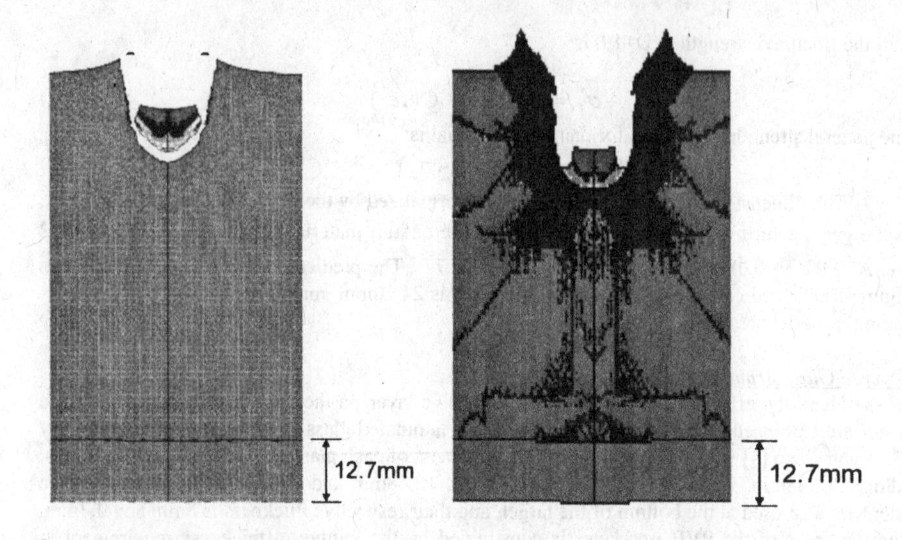

(a) 1-parameter prediction (b) Material #82 prediction

Figure 4. EPIC`06 predictions for monolithic glass armor using different material models

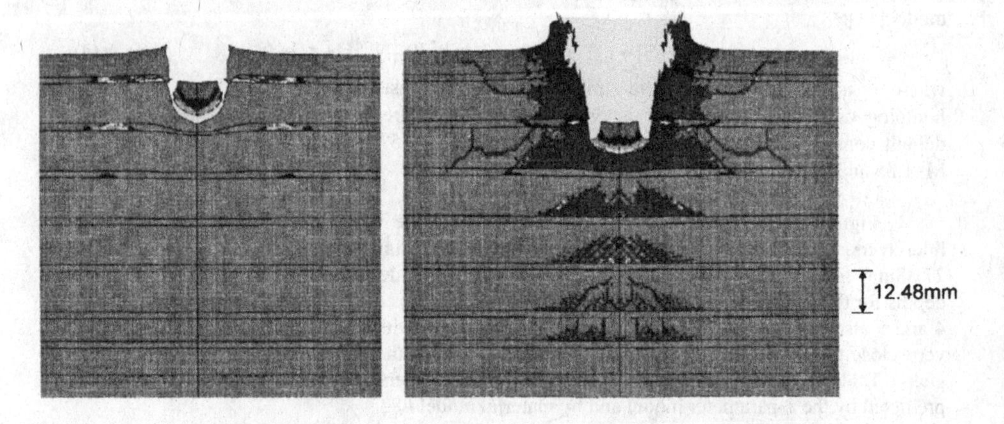

(a) 1-parameter prediction (b) Material #82 prediction

Figure 5. EPIC'06 prediction for layered glass armor

It is interesting to note that even though the static tests indicate the strength of intact soda lime glass is about 25% higher than that of the intact borosilicate glass [8], the same dynamic strength of 570MPa can be used for both soda lime glass and borosilicate to accurately predict penetration depth. Similar observations have also been made in Refs. 5 and 9 where the same critical strength value of 600MPa was used for both borosilicate glass and soda lime glass with continuum damage mechanics formulation to predict low speed impact damage.

DISCUSSIONS AND CONCLUSIONS

Results in Table 1 also indicate that the library material #82 for float glass over predicts the penetration depth for both the monolithic and the layered target. With the simple, constant flow stress model, EPIC'06 can predict the penetration depth of the layered glass armor with relatively good accuracy. This is true for both the gold rod penetration prediction on monolithic borosilicate [6] and the FSP impact simulation on laminated soda lime glass target. This indicates that the comminuted glass particles at the tip of the projectile flow like a viscous fluid and that the strength of the fractured glass follows the material behavior of a perfectly plastic flow without any strain hardening, regardless of glass type. These findings, if true, are important factors to be considered in penetration simulation [7, 10] and further studies on lower velocity impact should be conducted to examine the validity of these findings.

It should be mentioned that the predicted damage zone is much more concentrated around the projectile, as oppose to the multiple other fracture mechanisms as shown and discussed by Bless and Chen [4]. The damages away from the projectile impact zone are due to the stress waves generated by the impact shock, and this is the fundamental difference between the impact behaviors of brittle armor and ductile armor: Ductile armor has single penetration point surrounded by material undergone considerable amount of plastic deformation; For brittle armor, in addition to the penetration crater and the finely comminuted zone immediately adjacent to it, there are large regions of damaged/fractured zone in the target with various crack morphologies. These different, discrete damaged zones will influence the multi-hit capability of the brittle armor systems in different degrees, therefore a comprehensive glass damage model that can predict all the damage mechanisms should be developed for the brittle armor systems. The single-state 1-parameter model in EPIC'06 is sufficient in predicting the penetration depth and the damage around the impact crater, yet more development is needed to quantify and predict all the other damage mechanisms in brittle armor systems.

ACKNOWLEDGMENTS

Pacific Northwest National Laboratory is operated by Battelle for the U.S. Department of Energy under contract DE-AC06-76RL01830. Technical discussions with Mr. Tim Holmquist of Southwest Research Institute are also gratefully acknowledged.

FOOTNOTES

*Corresponding author, phone: 509 372 6489, fax 509 375 2604, email: xin.sun@pnl.gov

REFERENCES

[1]G.R. Johnson, S.R. Beissel, C.A. Gerlach, R.A. Stryk, T.J. Holmquist, A.A., Johnson, S.E. Ray and J.J. Arata, *User Instructions for the 2006 Version of the EPIC Code*, 2006.

[2]T. Behner and V. Hohler, Failure Kinetics of DEDF and Borosilicate Glass against Rod Impact, Presented at Progress Meeting EMI-SwRI-IRA-RDECOM at SwRI, San Antonio, 05-06, Oct. 2005.
[3]X. Nie, W. Chen, X. Sun and D.W. Templeton, Dynamic Failure of Borosilicate glass under Compression/Shear Loading: Experiments, Accepted for publication by *Journal of American Ceramics Society*, 2007.
[4]S. Bless and T. Chen, Impact Damage in Layered Transparent Armor, submitted to *International Journal of Impact Engineering*, 2007.
[5]X. Sun, W.N. Liu, W. Chen and D.W. Templeton, Modeling of Dynamic Failure of Borosilicate Glass Under Compression/Shear Loading, submitted to *International Journal of Impact Engineering*, 2007.
[6]G.R. Johnson and T.J. Holmquist, Determination of Simple Constitutive Models for Borosilicate Glass using Penetration-Velocity Data from Ballistic Experiments, Presented at the *15th APS Topical Conference on Shock Compression of Condensed Matter*, June 24-29, 2007, Hawaii.
[7]D. Grady, *Hydrodynamic Turbulence and the Catastrophic Comminution of Brittle Solids under Ballistic Penetration*, presented at the Glass Status Meeting, Purdue University, June 21-22, 2006.
[8]A. Wereszczak, K. Johanns and P. Patel, *Glass Characterization Update at ORNL*, presented at the Glass Status Meeting, Purdue University, June 21-22, 2006.
[9]X. Sun, M.A. Khaleel and R.W. Davies, Modeling of Stone-Impact Resistance of Monolithic Glass Ply Using Continuum Damage Mechanics, *International Journal of Damage Mechanics*, Vol. 14, April 2005. pp. 165-178.
[10]C.S. Alexander, T.J. Vogler, W.D. Reinhart and L.C. Chhabildas, *Summary of Glass Experiments at Sandia National Laboratories*, presented at Glass Review Meeting, Stanford Research Institute, Menlo Park, California, September 6-7, 2007.
[11]G.R. Johnson and W.H. Cook, A Constitutive Model and Data for Metals Subjected to Large Strains, High Strain Rates, and High Temperatures. In *Seventh International Symposium on Ballistics*, The Hague, The Netherlands, April 1983.

PHYSICS OF GLASS FAILURE DURING ROD PENETRATION

D. A. Shockey, D. Bergmannshoff, D. R. Curran, and J. W. Simons
SRI International
Menlo Park, CA 94025

ABSTRACT

The failure physics of glass when attacked by a projectile was investigated by examining the damage in glass target blocks that were partially penetrated by a steel rod. Fragment shapes and sizes in the tunnel surrounding embedded projectiles suggest penetration occurs by comminution of material at the nose of the advancing projectile and subsequent flow of the fine fragments out of the projectile path into more coarsely fragmented tunnel material. The observations and measurements are being used to develop a physics-based model that enables computational simulations of glass penetration scenarios and design of transparent armor.

INTRODUCTION

A goal of the Army is to reduce the weight of vehicle windows while meeting ballistic performance requirements. Current efforts to determine appropriate trade-offs between protection and weight are mostly experimental, where windows of different materials and structure are fabricated and then subjected to ballistic tests. Trade-offs can be examined more quickly and at lower expense with computational simulations, but the reliability and the detail of the computation results depend on how well the constitutive equations describe material failure behavior. Models deduced by fitting computational results to depths-of-penetration observed in ballistic tests do not always predict behavior in another impact scenario (see for example References 1-3), nor do they always compute certain features of interest, such as dwell time or the extent of damage. The ability of a window to survive a second impact, for example, requires that damage and strength loss associated with the first impact be modeled. For such purposes, a material model based on the failure physics of glass during penetration is needed.

To generate such a model, an understanding of the mechanism of penetration is required, as is a quantitative description of the damage. Our approach to obtain this information is to (1) perform ballistic experiments on glass targets at velocities such that the projectile penetrates only partially through the glass, (2) examine the targets post-test, and (3) characterize and quantify the damage.

PARTIAL PENETRATION EXPERIMENTS

Partial penetration of glass target blocks was achieved by accelerating steel rods to velocities in the 300 to 600 m/s range with a .458 magnum rifle. The projectile rods, which weighed 7.5 grams, were fitted in standard .458 shells with 3.4-gram sabots made of Delrin. The quantity of powder charge was adjusted to achieve desired velocities and a powder weight-projectile velocity calibration curve was determined in a series of gun firings.

Projectile rods 31.75 mm long and 6.35 mm in diameter with a hemispherical nose were machined from 6.35 mm maraging steel rod. The rods were heat soaked at 482°C for 6 hours, then slowly cooled to achieve a hardness of R_C 52. Mechanical properties were not measured, but in this heat treat condition the projectiles should have yield and ultimate tensile strengths of 1780 MPa and 1850 MPa, respectively.

Targets of soda lime glass 100x100x50 mm and borosilicate glass 195x195x75 mm were encased front and back with 9-mm-thick PMMA plates and on the sides with a PMMA "picture frame" to contain the fractured target after ingress of the projectile. The cover plates had a central hole to allow the impacting rod to hit only glass. Velocities of 300 to 600 m/s produced depths of penetration from 10 to 23 mm; therefore the nose of the arrested penetrator was 40 mm or more from the rear surface of the glass target. Several targets cracked through and the penetrator lay separated from the target pieces. The slightly tilted attitude of the arrested projectile suggested some yaw at impact. Figure 1 shows a target before and after impact.

Figure 1. Target assembly before and after impact.

After the impact, the encased targets with the embedded projectiles were carefully removed from the target mount to minimize shifting of the glass fragments, placed in the heated vacuum chamber on the rear plate (face plate up), and infiltrated with epoxy to hold the fragments in place and allow the fragmented target to be sectioned. A low-viscosity epoxy was poured into the front plate hole and the pouring chamber was evacuated overnight. The next day the target was inverted, the rear plate was removed, and the infiltration procedure was repeated for the rear target surface. After the targets were stabilized, the remaining confinement was removed and the targets were cut in two on a plane that included the penetration axis. Several targets were not stabilized with epoxy in order to collect loose fragments or compacted agglomerates for size distribution analysis.

DAMAGE OBSERVATIONS

The cross sections through the shot line revealed the uplifted target material near the impact site, the well-known cone and lateral cracks[4-9], and a concentric white frosted region around the projectile cavity, figure 2. The boundary between this white region and the more transparent region defines the tunnel. For the conditions of these experiments, the tunnel boundary (outer radius) is about 4 to 12 mm from the penetrator center line, or about 1 to 4 projectile radii.

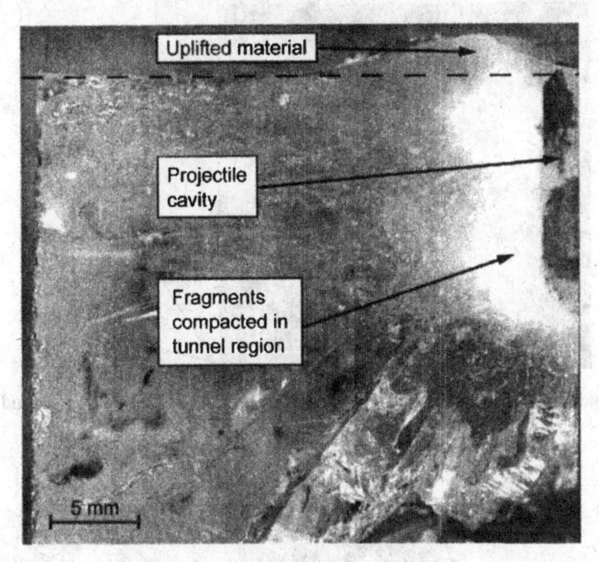

Figure 2. Section through target showing damage around impact site.

Similar cross sections near the nose of the arrested projectile but taken one projectile diameter from the shot line, Figure 3, show the cracking pattern and in-situ fragments, possibly illustrating how tunnel fragments form. Closely spaced (about 0.3-0.5 mm apart) cone cracks fan out in a divergent pattern and the long slender strips of glass between them are segmented by lateral cracks, producing rectangular fragments with aspect ratios ranging from 1 to 4.

Figure 4, a petrographic thin section parallel to the cross section in Figure 2, provides a clearer look at material in and around the tunnel (the circular white areas are air bubbles trapped during preparation of the thin sections). The transparent material at some distance from the projectile cavity in Figure 2 is heavily cracked and fragmented, but the fragments are in their original positions relative to each other. The white frosted region is more densely cracked and more finely fragmented. Moreover, the fragments have moved and rotated from their original positions. The boundary between the coarsely and finely fragmented regions is the boundary of the tunnel.

The densely cracked tunnel consists of colonies of in-situ fragments in a matrix of smaller, more randomly-oriented fragments. These colonies are nominally equiaxed with diameters up to 1 mm. In-situ fragments within the colonies are rectangular with aspect ratios of about 1 to 4, defined by perpendicular intersecting cone and lateral cracks. The orientation of the cone cracks within a colony with respect to the cone cracks outside the tunnel attests to rotation of the colony as the projectile penetrated.

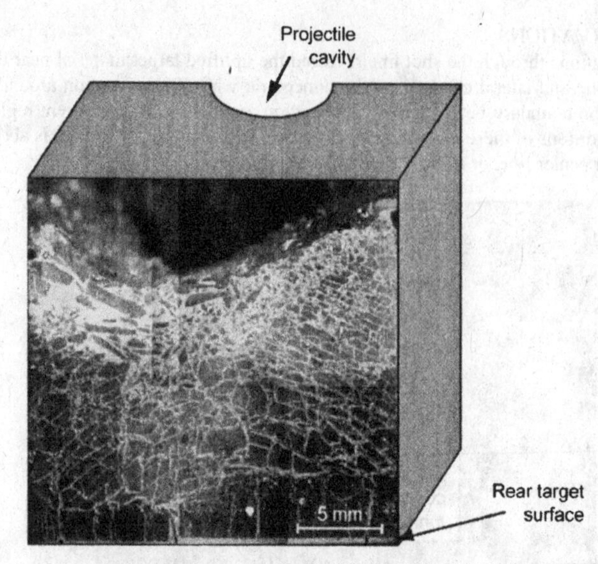

Figure 3. Off-center section showing in-situ fragments formed by intersecting cone and lateral cracks.

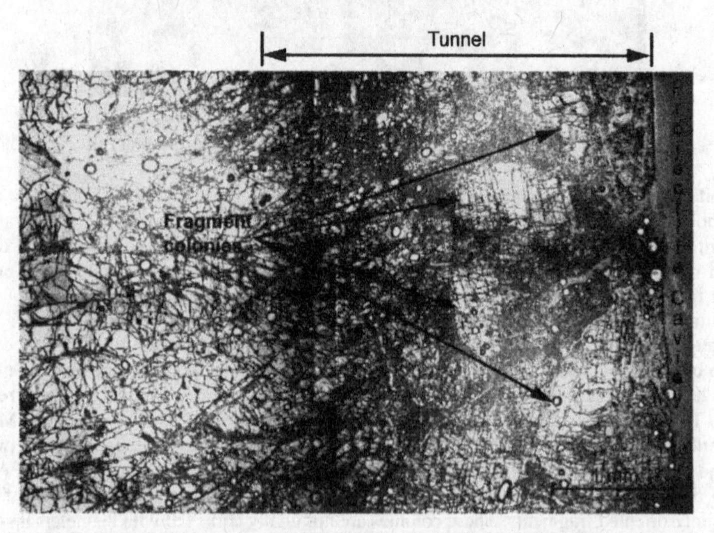

Figure 4. Thin petrographic section showing fragment colonies, individual fragment shapes, and fragment packing pattern in the tunnel surrounding the projectile.

Beneath the arrested projectile tip were closely-spaced cone cracks extending in the penetration direction, Figure 5. Little evidence of the Mescall zone* (MZ)[10,11], the highly stressed, finely comminuted volume of target directly beneath the nose of an advancing projectile, could be observed, because an arresting projectile does not produce high shear stresses and hence, does not produce the damage representative of a fast-moving projectile. However, small rounded fragments mixed in with the large angular fragments in the tunnel may be MZ fragments produced by a fast-moving projectile that have migrated away from the MZ.

Figure 5. Closely-spaced cone cracks beneath the nose of an arrested projectile.

The fragments in the tunnel are firmly compacted. For tests in which the glass targets cracked through, the projectile was not embedded, but was found lying in the test chamber with the tightly compressed tunnel material attached, Figure 6. This provided an opportunity to measure the fragment size distribution of tunnel material.

Figure 6. Agglomerated glass fragments attached to a projectile

* Named for John Mescall who first deduced its existence.[12,13]

Samples of this coherent glass powder agglomerate were taken from four locations — at the nose and half-shaft position of the penetrator and near the inside and outside boundaries of the tunnel. Triplicate measurements of fragment radius distribution in the range up to 120 microns were made at each location with a Horiba particle classifier. Figure 7 shows the size distribution of fragments having radii up to 30 microns. Fragment radii showed little variation with position in the tunnel. Figure 8 shows that the larger fragments are cube-like, bar-like, and plate-like with sharp edges and corners, having aspect ratios of 1 to 4. A few large (0.2 to 0.5 mm) fragment colonies, such as seen in Figure 4, were embedded in the agglomerate.

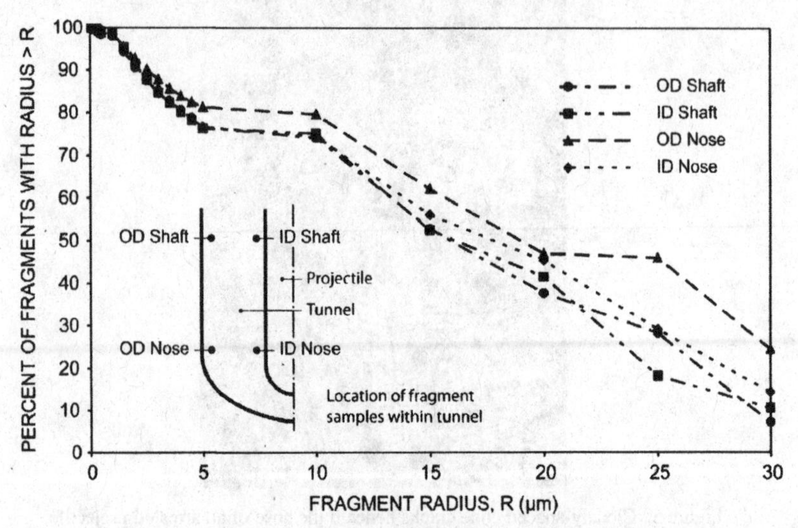

Figure 7. Size distribution of fragments in the < 30 μm at four locations within the tunnel of a partially penetrated target.

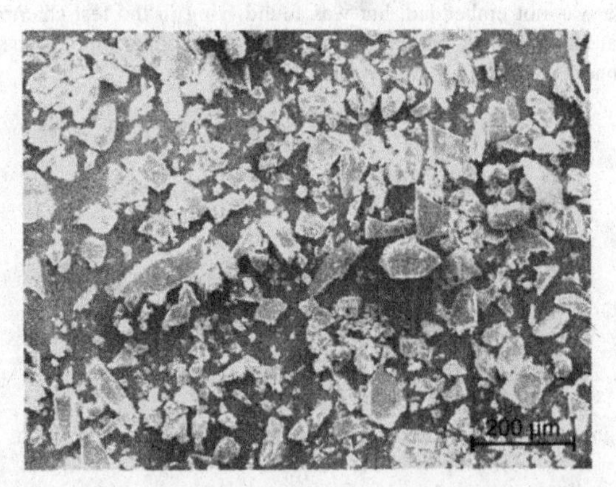

Figure 8. Glass fragments collected from a projectile tunnel.

PROJECTILE DAMAGE

Projectiles recovered after impact had axial grooves along the embedded portion of their shafts and a layer of glass attached to their noses, Figure 9(a). Patches of parallel grooves were evident on the projectile nose when the adherent glass layer was removed, Figure 9(b). The grooves attest to the abrasive effect of glass fragments as the projectile advances in the target.

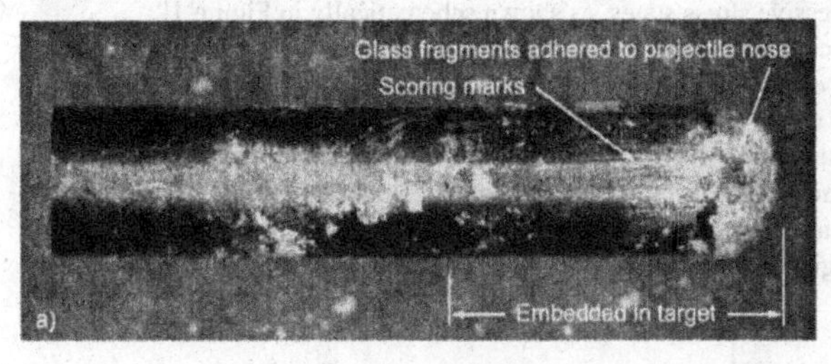

Figure 9. Scoring marks on recovered projectile (a) and on nose (b)

Close examination of the glass layer on the projectile nose, Figure 10(a) shows smeared areas that have cracked, suggesting that the high pressure and the interfragment friction generate enough heat to soften and perhaps melt glass fragments. Surfaces that were in contact with the projectile shaft during penetration exhibit small glass globules and lines of globules, Figure 10(b). Petrographic thin sections such as shown in Figure 4 also suggest that a thin layer of fused glass may exist next to the projectile.

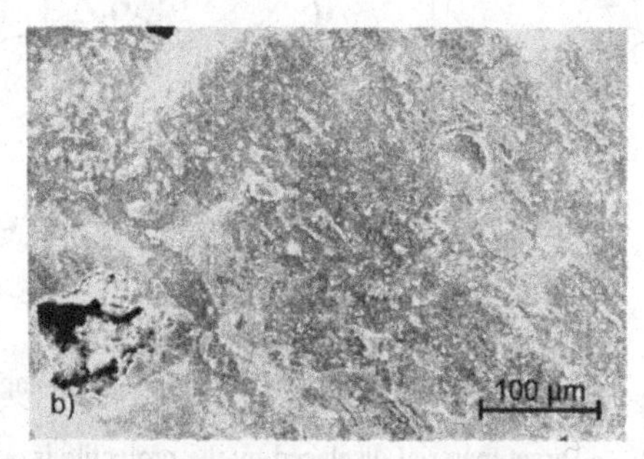

Figure 10. Melting evidence on the glass layer adhering to the projectile nose (a) and on a glass surface that was adjacent to the projectile shaft (b).

The surface of the fragment agglomerate in contact with the projectile shaft in Figure 6 was blackened and coherent, suggesting frictional heating and chemical interaction between projectile and target. Chemical analysis of this surface showed iron, nickel, titanium, and molybdenum, the main constituents of the maraging steel projectile and in approximately the same percentage. Analysis of the agglomerate further from the projectile showed none of these elements. Thus, the projectile was slightly eroded thermally, chemically, and mechanically as it penetrated the glass target.

DISCUSSION

These observations suggest that penetration occurs by comminution of glass in the Mescall zone ahead of the projectile nose and the extrusion of the fine glass fragments out of the projectile path and into a more coarsely fragmented "tunnel" region surrounding the projectile. The cracking and fragment patterns are consistent with three distinct stress zones around the projectile, Figure 11. A material particle ahead of the projectile is loaded, damaged, and displaced in three successive steps under consecutive tensile, shear, and compressive stress states, as shown schematically in Figure 12.

The material initially experiences tension, and acquires closely-spaced cone cracks running at slight radial angles to the penetration direction, Zone 1. Subsequent lateral cracks break up the material between adjacent cone cracks. Next, as the projectile moves closer, a local volume (about the size of the projectile nose) of this cracked material is overrun by a low-confinement field of high shear and is comminuted into fine fragments, Zone 2. Thirdly, the projectile reaches the comminuted material and imposes high pressure, Zone 3, extruding the comminuted material to the sides of the projectile nose and into the cracked and coarsely fragmented tunnel.

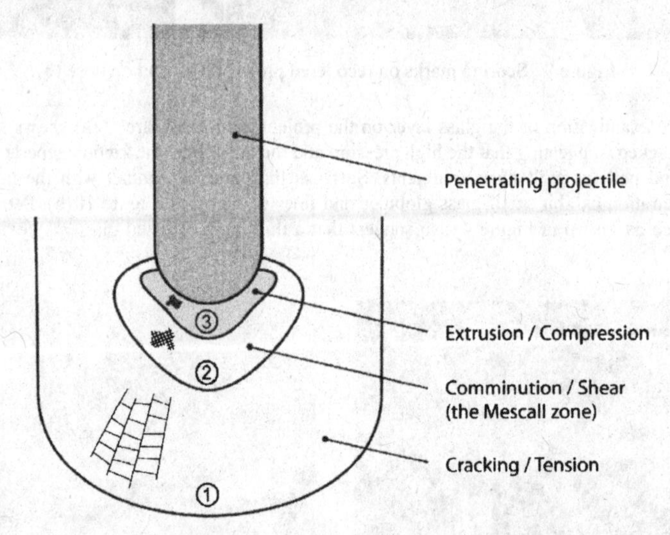

Figure 11. Stress conditions and damage activity ahead of a penetrating projectile

Target material displaced by the projectile is accommodated in the early stages of penetration by ejection of target fragments and uplift at the impact surface. At later stages the flow of fragments is probably primarily in the radial direction, a flow accommodated by favorably oriented macrocracks in the tunnel, which are wedged apart to provide paths for migrating MZ material. Flow may be facilitated by frictional heat generated during penetration, which may soften or melt glass fragments, and by the slight rotation of colonies of cracked tunnel material.

Since target material must flow out of the path of the projectile for the projectile to advance, microcracking in the MZ is the key underlying mechanism governing penetration. The intensity of MZ microcracking dictates MZ fragment size and shape, which in turn dictate flow behavior. Thus, microcrack numbers, sizes, spacings, and orientations in the MZ are necessary data for failure physics models of penetration. Although posttest damage is not representative of dynamic processes occurring ahead of a projectile, quantitative estimates of microcrack size and density in the MZ may be obtained from the sizes and shapes of MZ fragments in the tunnel region. MZ fragments should be distinguishable from fragments that formed in the tunnel. They should be smaller than tunnel fragments, because they form under higher compression and shear conditions. Moreover, they should be less angular than tunnel fragments, since they probably rotate and rub edges off as they flow out of the MZ and, further, may have softened under compressional and frictional heat.

Future work is aimed at modeling MZ microfracture activity, a task that requires identifying MZ fragments among the fragments in the tunnel and measuring their size distribution. The MZ fragment size distribution will be analyzed to obtain estimates of MZ microcrack numbers and sizes. Laboratory tests will be performed to measure frictional flow properties of fragment beds, i.e., shear strength as a function of pressure. These data will be used to develop a computational model of rod penetration into glass. The framework of this fracture-physics-based model is presented in a paper which follows[14].

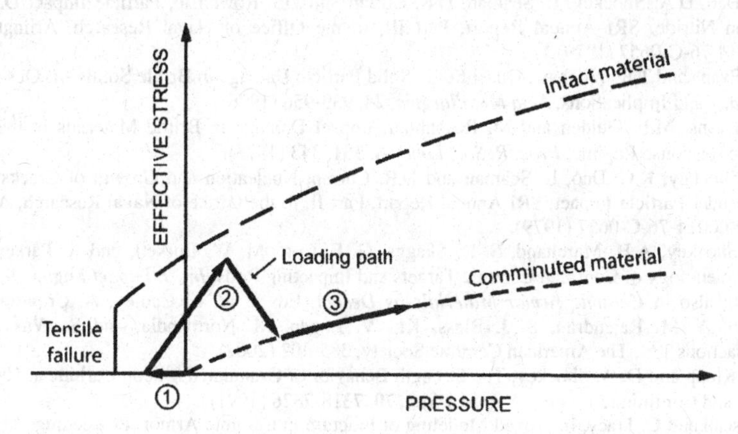

Figure 12. Strength of intact and comminuted glass and the stress path experienced by glass during penetration.

SUMMARY

Post-test examination of glass targets partially penetrated by a projectile rod suggests that target material in advance of the projectile is densely fractured into micron-sized fragments, which are extruded into more coarsely cracked target material around the projectile. Glass fragments score the nose and shaft of the projectile as it penetrates, generating frictional heat that softens and perhaps melts the glass. These observations and the data being generated on fragment sizes, shapes, and flow behavior are the basis of a computational model presented later in this conference.[14]

ACKNOWLEDGMENTS

This work was funded by the U.S. Tank Automotive Research, Development, and Engineering Center under subcontract from Southwest Research Institute. The authors are grateful to D. W. Templeton of TARDEC and C. E. Anderson, Jr. of SwRI and their teams for their interest and support.

REFERENCES

[1] S.Chocron, C.E.Anderson, Jr., and A.E.Nicholls, Constitutive Model for Borosilicate Glass and Application to Long-Rod Penetration. 23rd Int. Symp. Ballistics, **2**, 1073-1081, Gráficas Couche, Madrid, Spain (2007).

[2] S.Chocron, K.A.Dannemann, J.D.Walker, A.E.Nicholls, and C.E.Anderson, Jr. Constitutive Model for Damaged Borosilicate Glass under Confinement, *J. Am Cer. Soc.*, 90(8), 2549-2555 (2007).

[3] S.Chocron, K.A.Dannemann, J.D.Walker, A.E.Nicholls, and C.E.Anderson, Jr., Analytic Model of the Confined Compression Test Used to Characterize Brittle Materials, *J. Appl. Mech.*, accepted for publication (2007).

[4] B.Lawn and T.R. Wilshaw, Review of Indentation Fracture: Principles and Applications, *J. Mater. Sci.* 10, 1049-1081 (1975).

[5] A.G. Evans, Impact Damage in Ceramics, p. 302 in *Fracture Mechanics of Ceramics,* Vol. 3, Edited by R.C. Bradt, D.P.H. Hasselmann and F.F. Lange, Plenum Press, New York (1978).

[6] K.C. Dao, D.A. Shockey, L. Seaman, D.R. Curran and D.J. Rowcliffe, Particle Impact Damage in Silicon Nitride, SRI Annual Report, Part III, to the Office of Naval Research, Arlington, VA, N00014-76-C-0657 (1979).

[7] A.G. Evans and T.R. Wilshaw, Quasi-Static Solid Particle Damage in Brittle Solids - I. Observations, Analyses and Implications, *Acta Metallurgica*, **24**, 939-956 (1976).

[8] A.G. Evans, M.E. Gulden and M. Rosenblatt, Impact Damage in Brittle Materials in the Elastic-Plastic Response Régime, *Proc. R. Soc. Lond.* A, **361**, 343 (1978).

[9] D.A. Shockey, K.C. Dao, L. Seaman and D.R. Curran, Nucleation and Growth of Cracks in CVD ZnS Under Particle Impact, SRI Annual Report, Part II, to the Office of Naval Research, Arlington, VA, N00014-76-C-0657 (1979).

[10] D.A. Shockey, A.H. Marchand, S. R. Skaggs, G. E. Cort, M. W. Burkett, and R. Parker, Failure Phenomenology of Confined Ceramic Targets and Impacting Rods, *Int. J. Impact Engng.*, **9**, 263-275 (1990); also in *Ceramic Armor Materials by Design*, Eds J. W. McCauley, A. Crowson, W. A. Gooch, A. M. Rajendran, S. J. Bless, KL. V. Logan, M. Normandia, and S. Wax, Ceramic Transactions 134, The American Ceramic Society, 385-402 (2002).

[11] R.W. Klopp and D. A. Shockey, The Strength Behavior of Granulated Silicon Carbide at High Strain Rates and Confining Pressure, *J. Appl. Phys.*, **70**, 7318-7326 (1991).

[12] J. Mescall and C. Tracy, Improved Modeling of Fracture in Ceramic Armor, Proceedings of the 1986 Army Science Conference, U.S. Military Academy, West Point, June 17-20 (1986).

[13] J. Mescall and V. Weiss, Materials Behavior Under High Stress and Ultra-high Loading Rates-Part II, Proceedings of the 29th Sagamore Army Conference, Army Materials and Mechanics Research Center, Watertown, MA (1984).

[14] D.R. Curran, D. A. Shockey, and J. W. Simons, Mesomechanical constitutive relations for glass and ceramic armor, presented at the 32nd International Conference and Exposition on Advanced Ceramics and Composites (ICACC), Daytona Beach, Florida, Jan. 27 – Feb. 1 (2008).

ADHESIVE BOND EVALUATION IN LAMINATED SAFETY GLASS USING GUIDED WAVE ATTENUATION MEASUREMENTS

S. Hou and H. Reis
Department of Industrial and Enterprise Systems Engineering
University of Illinois at Urbana-Champaign
104 South Mathews Avenue
Urbana, IL, 61801

ABSTRACT

Laminated safety glass samples with different levels of adhesive bond strength were manufactured and tested using mechanical guided waves. The results were then compared with those obtained using the commonly used destructive testing method, i.e., the pummel test method. The imperfect interfaces between the plastic interlayer and the two adjacent glass plates were modeled using a bed of longitudinal and shear springs. The spring constants were estimated using fracture mechanics concepts in conjunction with surface analysis using atomic force microscopy. The guided wave theoretical attenuation-based predictions of adhesion levels (obtained using this multilayered model) were validated using experimental attenuation measurements. Results show that this approach is useful in the nondestructive assessment of adhesive bond strength in laminated safety glass.

INTRODUCTION

Laminated safety glass[1], see Figure 1 and Table I, is widely used in windshields for automotive and other vehicle applications; architecture applications, as in windows for skyscrapers; and bullet proof glass for military and security uses. Laminated safety glass consists of two or more glass sheets bonded by an interlayer of transparent, adherent plastic. The glass sheets in windshield laminates are made from either plate glass, float glass, tempered glass and are approximately 0.1 inches (2.52 mm) thick[2,3]. The plastic interlayer is made of a thermoplastic material, such as polyvinyl butyral (PVB), and thicknesses of 0.015 inches (0.381 mm) or more are commonly used[4].

Laminated safety glass for windshield applications is manufactured to improve the impact and penetration resistance against flying objects and to minimize the danger of flying glass after impact. Penetration is prevented mainly by absorbing the kinetic energy through stretching the plastic interlayer, delamination between the plastic interlayer and the two adjacent glass plates, and by fracture of the adjacent glass plates. The plastic interlayer must delaminate partially to allow

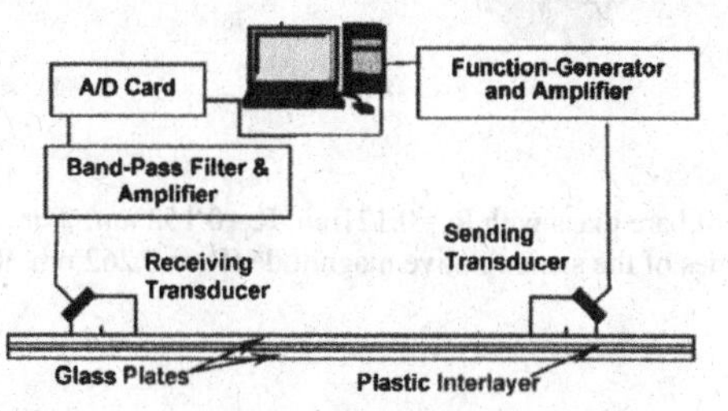

Figure 1. Schematic diagram of guided wave measuring system to assess levels of adhesion in safety glass.

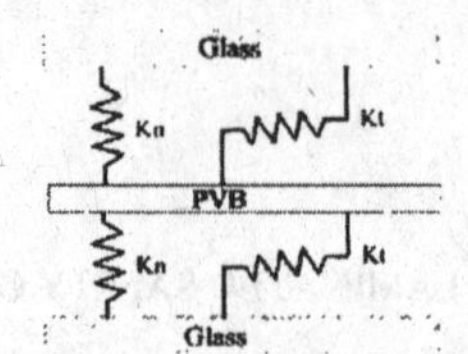

Figure 2. Spring model for imperfect interfaces of laminated safety glass.

consumption of large amounts of energy. Yet, to minimize the danger of flying glass after impact, delamination must not be excessive. Therefore, the best impact performance requires an optimal level of adhesion between the plastic interlayer and the two adjacent glass plates[5,6]. A good review of the impact and penetration resistance of laminated safety glass is given by Huntsberger[6].

There exist several destructive test methods to evaluate the overall impact performance of laminated safety glass. These methods, include the 90° peel test[7,8], the compressive shear test[7], and the impact test method which is a commonly used method to rate overall laminate performance as described in safety standards test procedures[4,8,9]. However, the most widely used destructive test method for quality control monitoring of laminated safety glass is the destructive 0° Fahrenheit pummel test[4].

In the 0° Fahrenheit pummel test[4], a 12 inch by 12 inch (30.5 cm by 30.5 cm) is cut into four nominal 6 inch by 6 inch (15.3 inch by 15.3 inch) specimens. These specimens are then conditioned at 0° Fahrenheit (-18 °C) for at least one hour. The laminate is then pounded progressively in ½ inch (12.7 mm) increments along the bottom ¾ inch (19.1 mm) of the laminate. When the bottom edge has been completely pulverized, the next ¾ inch (12.7 mm) is pulverized. All the smooth glass in the pummeled section must be completely pulverized, and all loose glass removed. The pummeled samples are allowed to reach room temperature and the condensed moisture to evaporate before grading begins. The specimens are compared to standards, which are pummeled specimens representing the range of possible adhesive bond strength. There are ten standards rated from 1 to 10, with 1 corresponding to very low and 10 to very high adhesive bond strength. The specimens with lower adhesive bond strength will have less pulverized glass remaining, and more of the plastic

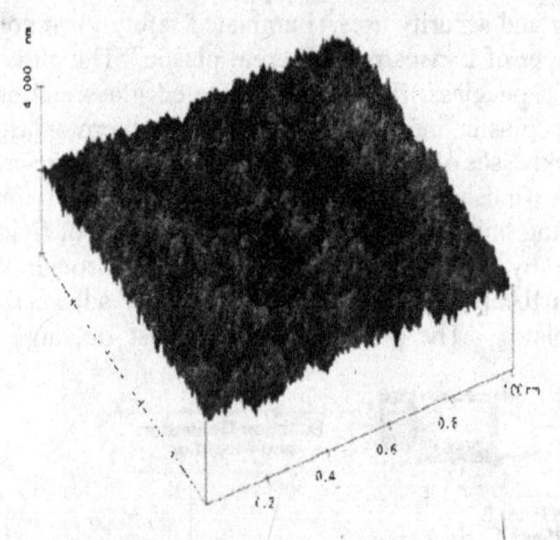

Figure 3. AFM height image of bare glass with R_a=0.121nm, R_q=0.152 nm. For comparison, PVB layer after lamination has values of the same relative magnitude (R_a = 0.262 nm, R_q = 0.330 nm).

Table I. Materials properties for the laminated safety glass.

	Thickness (mm)	Density g/cm³	E GPa	Poisson Ratio	Long. Attenuation np/m	Shear Attenuation np/m
Glass	2.54	2.5	72	0.25	0	0
PVB	0.762	1.1	3.9	0.34	8.5106	43.478

Table II. Spring stiffness values for different adhesion levels.

Pummel Number	Surface Bearing Ratio	Crack Ratio (a/b)	Normal Spring Stifness (K_n) (N/m)/m²	Shear Spring Stifness (K_t) (N/m)/m²
3	35.1%	0.649	6.8785E+11	2.4091E+11
4	45.8%	0.542	8.9754E+11	3.1436E+11
5	55.0%	0.450	1.0778E+12	3.7750E+11
6	76.2%	0.238	1.4933E+12	5.2301E+11
8	86.8%	0.132	1.7010E+12	5.9577E+11

interlayer will be exposed. The amount of exposed interlayer can be determined by holding the specimens in fluorescent light and estimating how much light is reflected by the plastic. More exposed plastic means more light reflected, which in turn means weaker adhesive bond strength. The specimens are then given a rating between 1 to 10 depending on which of the standards most nearly reflects the same amount of light. The ten pummel test standards are calibrated using the impact test method[4].

The fact that the above-mentioned test methods are intrinsically destructive precludes their use as viable on-line quality control techniques. In addition, these tests raise questions as to their practicality as adhesive bond strength testing procedures. Because of all these drawbacks and because adhesion plays a key role in the performance of safety glass, the need for nondestructive test methods for the evaluation/characterization of adhesive bond strength in laminated safety glass is apparent. Recently, Hou and Reis[10], used energy velocities to estimate adhesive bond quality in laminated safety glass. The purpose of this study is to investigate the use of attenuation measurements to evaluate adhesive bond quality in laminated safety glass.

SURFACE CHARACTERIZATION

The interfaces between the plastic interlayer and the two adjacent glass plates are modeled as imperfect interfaces via the use of a bed of longitudinal and shear springs[10-13], as illustrated in Figure 2. The stiffness of these springs were estimated by taking advantage of fracture mechanics concepts[14-16] and by the surface characterization of the glass plates and the PVB plastic interlayer[17]. Acoustic waves have already been used to study interfaces[18-33]. A review is provided by Hou and

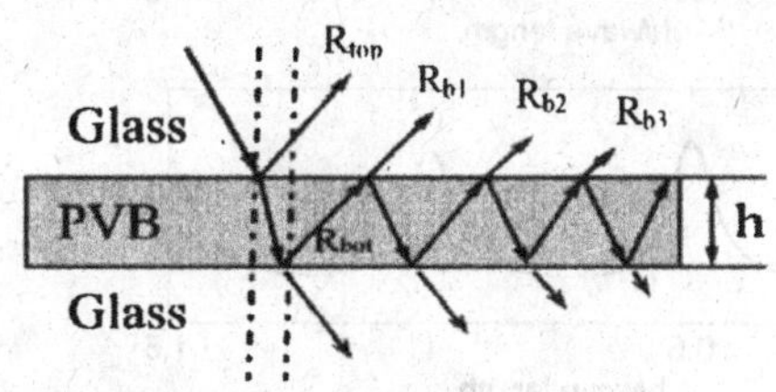

Figure 4. Reflections and transmissions of a plane wave for a model made of a plastic interlayer perfectly bonded to two adjacent glass half-spaces.

Reis[10].

Surface roughness is important in surface characterization[17]. Two of the most important parameters are the average roughness (R_a) and the root-mean-square of roughness (R_q). Average roughness is defined as the area between the surface profile and its mean base-line and it is defined as:

$$R_a = \frac{1}{L}\int_0^L |h(x)| dx \qquad (1)$$

where h(x) is the surface height. The root mean square roughness is defined as the square root of the average of the square of vertical surface deviation from a reference base-line, and it can be expressed as:

$$R_q = \sqrt{\frac{1}{L}\int_0^L h(x)^2 dx} \qquad (2)$$

A Sloan Dektak3 ST stylus surface profilometer was used for measurement of the surface topography of the PVB surface before lamination. The PVB surface topography before lamination is very rough with surface variations in the order of micrometers and as a result it is not transparent before being laminated between two adjacent glass layers. The profilometer operates by lightly dragging a sharp diamond stylus over the surface of the substrate and recording the vertical profile of the surface.

Because of the plastic deformation of the highest asperities during the lamination process, the roughness of the PVB layer is significantly reduced after lamination. After lamination the surface roughness of the plastic interlayer is close to the glass surface roughness of approximately 0.2 nm. Atomic force microscope (AFM) was used to measure the roughness of glass and PVB layers after lamination. AFM is a high-resolution imaging technique that can resolve features as small as an

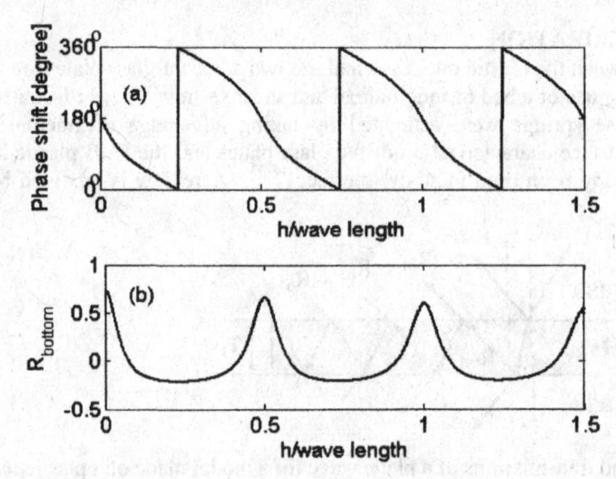

Figure 5. (a) Phase shift between top-reflection and bottom-reflection as a function of the ratio of the interlayer thickness to wavelength; (b) Total bottom-reflection coefficient as a function of the ratio of the interlayer thickness to wavelength.

atomic lattice. Figure 3 shows the three-dimensional surface profile for the surfaces of the two adjacent glass plates. The surface of the PVB layer is significantly altered by the lamination process; it reduced the PVB surface asperity height from micrometer level to nanometer level, the same order of magnitude as the glass plates.

The "actual" area of contact is known as the bearing area and may be approximately obtained from a surface profile or surface map[17]. A bearing area analysis consists in studying the distribution of surface heights for a sample revealing how much surface is above or below an arbitrarily chosen height (considered as a reference height) based on a statistical analysis. Initially, when two surfaces come into contact under load, the contact usually occurs only at a few asperities (the higher asperities). As the two surfaces move closer together under increasing normal load, these asperities will deform plastically under increasing contact stresses and a larger number of asperities will bond together. Higher adhesion levels occur as a result of this increase in contact area or bearing area[17].

Higher bearing ratios correspond to higher adhesion levels and to a corresponding lower crack ratio a/b where a and b are estimations of the delaminated area and the total interface area, respectively[10,14-16]. The crack ratio (a/b) can be expressed[10,14-16] as a function of the bearing ratio α as:

$$a/b = 1-\alpha \tag{3}$$

A proper reference height value must be selected to calculate the bearing area ratio. Using all the atomic force microscopy surface images (corresponding to all the laminates with different levels of adhesion), relationships between the bearing area and bearing depth were generated for laminates with different pummel ratios. A height of 1.0 nm was chosen experimentally to estimate the bearing area, which is related to the crack ratio (a/b) for each laminate. Then, using concepts from fracture mechanics[10,14,15], the longitudinal and shear spring stiffness were estimated as shown in Table II.

A SIMPLE MULTI-LAYERED MODEL

Simple layered models have been studied to shed light into the reflection and transmission coefficients of plane waves in multilayered media[26-29]. Simonetti[26] investigated Lamb wave propagation in elastic plates coated with viscoelastic materials, and Simonetti and Cawley[28] studied the nature of shear horizontal wave propagation in plates coated with viscoelastic materials. These

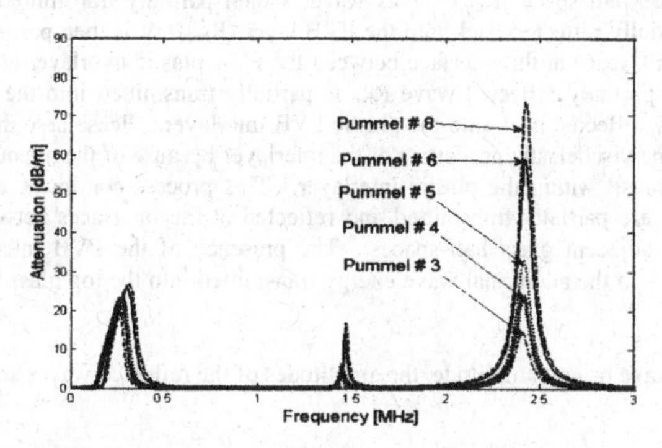

Figure 6. Attenuation dispersion curves of S_1 mode for different level of adhesion, i.e., pummel test ratings.

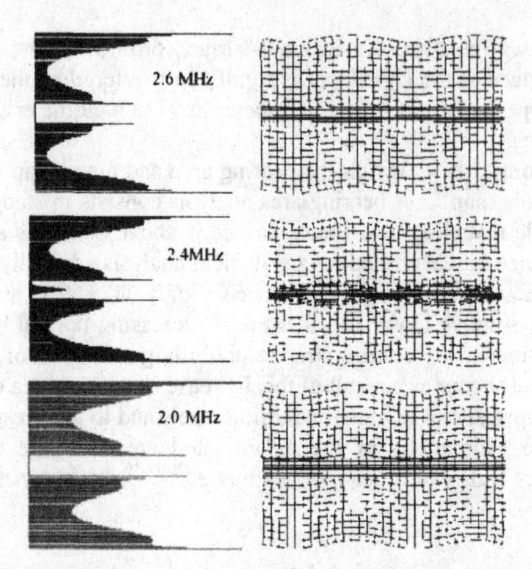

Figure 7. Relative power flow profile and mode shape distribution through the laminate thickness for the S_1 mode at different frequencies for laminates with a pummel rating of eight.

models help to explain the attenuation peaks observed in multi-layered structures (three layers for laminated safety glass), attributing the result to resonance phenomena that cause maximum energy transfer from the two surrounding adjacent glass plates into the viscoelastic PVB layer with possible constructive and destructive interference at the interfaces.

A simple three-layered system is now presented and discussed. The simple system consists of a viscoelastic (PVB) interlayer perfectly bonded to two adjacent elastic glass half-spaces as shown in Figure 4. This simple model exhibits periodic attenuation peaks as a function of frequency as shown in Figure 5, which are explained by the transmission and reflection of a plane acoustic wave at the interfaces between the plastic PVB interlayer and the two adjacent glass half-spaces. Assuming a plane wave incident from the top glass half-space, it is partially reflected at the interface between the PVB layer and top glass half-space (R_{top}). This wave is then partially transmitted into the PVB interlayer, and then partially reflected back into the PVB layer (R_{bot}). It is then partially transmitted into the bottom glass half-space at the interface between the PVB plastic interlayer and bottom glass half-space. Again, the partially reflected wave R_{bot} is partially transmitted into the top glass half-space (R_{b1}) and partially reflected back into the plastic PVB interlayer. Please note that (R_{b1}) carries information regarding the viscoelastic properties of the interlayer because of the attenuation the wave encountered while in transit within the plastic interlayer. This process continues, as illustrated in Figure 4, where waves are partially transmitted and reflected at the interfaces between the plastic interlayer and the two adjacent glass half-spaces. The presence of the PVB interlayer of finite thickness "h" contributes to the additional wave energy transmitted into the top glass half-space (i.e., R_{b1}, R_{b2}, R_{b3}....).

Assuming an incident wave of unit amplitude, the amplitudes of the reflected waves are:

$$R_{top} = R_{GP}$$
$$R_{b1} = T_{GP} \cdot R_{PG} \cdot T_{PG} \cdot e^{2ik_P h}$$

$$R_{b2} = T_{GP} \cdot R_{PG} \cdot R_{PG} \cdot R_{PG} \cdot T_{PG} \cdot e^{4ik_p h} = T_{GP} \cdot R_{PG}^3 \cdot T_{PG} \cdot e^{4ik_p h}$$
$$R_{b3} = T_{GP} \cdot R_{PG}^5 \cdot T_{PG} \cdot e^{6ik_p h}$$
$$\vdots \tag{4}$$

where "h" is the thickness of the PVB layer. k_p is the complex wave number, and R_{GP} and T_{GP} are the reflection and transmission coefficients, respectively. The subscript GP means that the wave travels from the glass layer into the PVB layer, and vice versa for the subscript PG. The reflection and transmission coefficients can then be expressed as:

$$R_{PG} = \frac{Z_{glass} - Z_{pvb}}{Z_{glass} + Z_{pvb}}, \qquad T_{PG} = 1 - R_{GP}, \qquad R_{GP} = -R_{PG}, \qquad T_{GP} = 1 + R_{GP} \tag{5}$$

where Z_{glass} and Z_{pvb} are the acoustic impedances for the glass plates and the PVB interlayer, respectively. The acoustic impedance is defined as the product of the material density and bulk velocity, which can be obtained by noting that the complex velocity is given by[28]:

$$C_{comp} = \frac{C_{bulk}}{1 - i(\varsigma / 2\pi)} \tag{6}$$

where C_{bulk} and ς is the bulk velocity and the bulk attenuation in Nepers per wavelength, respectively, of the plastic PVB interlayer.

For sake of simplicity in illustrating how the peaks shown in Figure 5b occur, the bulk velocity and bulk attenuation per wavelength are assumed to be constants in this model[28]. The main interest is focused on those waves that travel through the PVB interlayer, since the energy absorbed by the viscoelastic PVB interlayer affects the energy of those waves. Therefore, summing all the waves transmitted into the top glass half-space (i.e., R_{b1}, R_{b2}, R_{b3}...), as illustrated in Figure 4, leads to the following total "bottom" reflected coefficient caused by the presence of the finite thickness viscoelastic interlayer[27]:

$$\begin{aligned} R_{bottom} &= R_{b1} + R_{b2} + R_{b3} + \cdots \\ &= T_{GP} \cdot R_{PG} \cdot T_{PG} \cdot e^{2ik_p h} + T_{GP} \cdot R_{PG}^3 \cdot T_{PG} \cdot e^{4ik_p h} + T_{GP} \cdot R_{PG}^5 \cdot T_{PG} \cdot e^{6ik_p h} + \cdots \\ &= T_{GP} \cdot R_{PG} \cdot T_{PG} \cdot e^{2ik_p h} \sum_{n=0}^{\infty} [R_{PG}^2 e^{2ik_p h}]^n \end{aligned} \tag{7}$$

For an infinite geometric progression, Equation (4) can be expressed as:

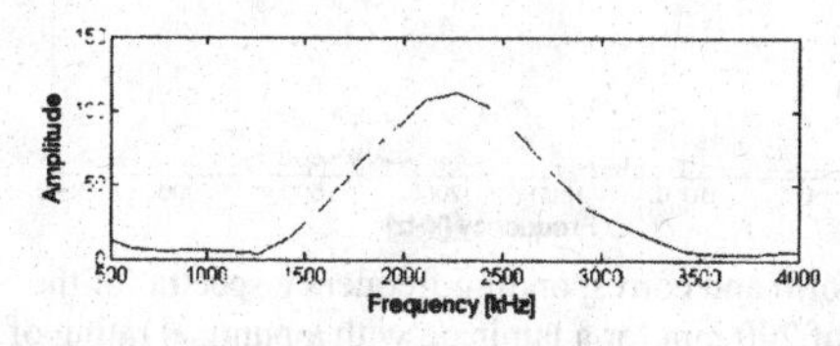

Figure 8. Transducer sensitivity response (test carried out with transducers face-to-face).

$$R_{bottom} = \frac{T_{GP} \cdot R_{PG} \cdot T_{PG} \cdot e^{2ik_p h}}{1 - R_{PG}^2 \cdot e^{2ik_p h}} \quad (8)$$

Figure 5 shows the phase shift between the top and the bottom reflections (R_{top} and R_{bottom}) and the coefficient of the bottom reflections as a function of the ratio of the PVB thickness to wavelength. In Figure 5a. the phase shift of the total bottom reflections is dominated by the R_{b1} (because the amplitude of R_{b2}, R_{b3} .... are far smaller than R_{b1}). It should be noted that the periodic attenuation peaks [(h/wave length) = 0, 0.5, 1,...] occur when the top and bottom reflections are opposite in phase, causing destructive interference between the reflections. This result is in agreement with Simonetti and Cawley[28] shear horizontal wave two-layer model study.

Please note that in this simple model, perfect bonding is assumed between the plastic interface and the two adjacent glass half-spaces and the bulk attenuation per wavelength is assumed to be constant. As a consequence, the model results can not be directly compared with those shown in Figure 6 for laminated safety glass, where these assumptions are not valid. However. Figure 5 helps explaining the periodic peaks in the attenuation curves associated with the S_1 mode in multi-layered safety glass as caused by the destructive interference phenomena that occurs at the interfaces[28].

GUIDED WAVE ANALYSIS AND EXPRIMENTAL RESULTS

The experimental set-up to measure attenuation is shown in Figure 1, where both the sending and receiving transducers are mounted on the same side of the test specimen. Five laminated safety glass samples were manufactured with controlled adhesive bond strength between the PVB layer and the two adjacent glass plates by Solutia Inc. These laminates had a pummel rating of 3, 4, 5, 6, and 8, see Tables I and II. Each laminate had the same adhesion level at the interfaces between the PVB interlayer and the two adjacent glass plates. A computer program named "DISPERSE," which is based upon the global matrix method[24,30,31], was used to calculate the dispersion curves[30]. The imperfect interfaces were modeled as a layer of springs with zero thickness and zero mass[14,22,24,30]. The material properties and thickness for the glass and PVB layers are shown in Tables I and II.

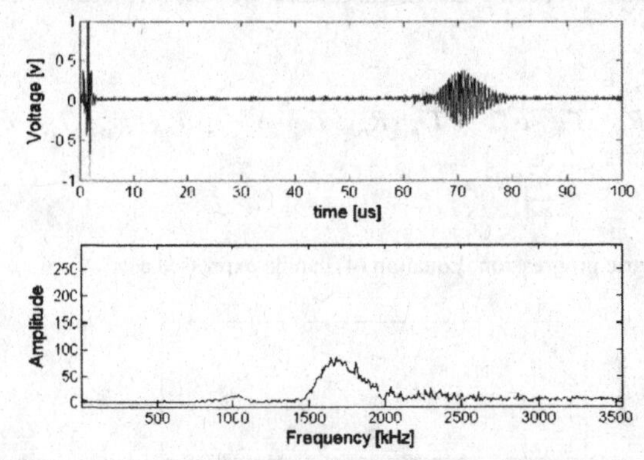

Figure 9. Time domain waveform and corresponding frequency spectra for the S_1 mode, with a transducer separation distance of 200 mm for a laminate with a pummel rating of 4.

Attenuation Dispersion Curves

Attenuation curves can characterize damping of a system, and they allow the examination of the decay properties of guided wave modes. Figure 6 shows the attenuation curves for the S_1 mode for laminated specimens with different adhesion levels, i.e., different pummel ratings. Figure 6 shows that the attenuation for the S_1 mode exhibits periodic peaks in certain frequency bands such as from 100 to 400 kHz and from 2.2 to 2.5 MHz.

The strain energy rate that flows through the laminated safety glass specimens and the location of energy transmission through the laminate thickness as well as the mode shapes were evaluated to shed light into the phenomena observed in Figure 6 at the frequency range of 2.0 to 2.6 MHz. Figure 7 shows the power flow and mode shape distribution through the thickness of a laminate with an adhesion level corresponding to the pummel rating of eight at different frequencies for the S_1 mode. At relatively lower frequencies (2.0 MHz), the power flow is concentrated in the two adjacent glass layers and there is very little power flowing in the PVB layer. By increasing the frequency, the energy is transferred from the glass layers to the PVB layer. At around 2.4 MHz, the energy flow in the PVB layer is at its peak. Above 2.4 MHz, the energy in the PVB layer decreases, until most of the wave energy is again concentrated in two adjacent glass layers at a frequency of 2.6 MHz as shown in Figure 7. Because of the relatively high attenuation of the PVB layer material (as compared to glass), the laminated safety glass specimens reach its highest attenuation at 2.4 MHz. An excellent physical description of this phenomena is given by Bernard, Lowe & Deschamps[29].

In laminated safety glass, with finite thickness glass plates and PVB plastic interlayer, the energy transferred from the glass layers into the viscoelastic interlayer reaches a local maximum at the resonance frequencies as illustrated in the power flow profiles shown in Figure 7. In Figure 6, it is also observed that higher adhesion levels, which correspond to higher values for spring stiffness (see Table II), lead to higher attenuation values. This occurs because springs with higher stiffness values transfer, i.e., pump, more energy into the PVB viscoelastic layer, which leads to further increase in energy dissipation in the PVB layer, i.e., in the viscoelastic layer.

Experimental Results

To validate the predicted attenuation values based upon the spring model and the AFM measurements, experimental measurements were also carried out using laminated safety glass

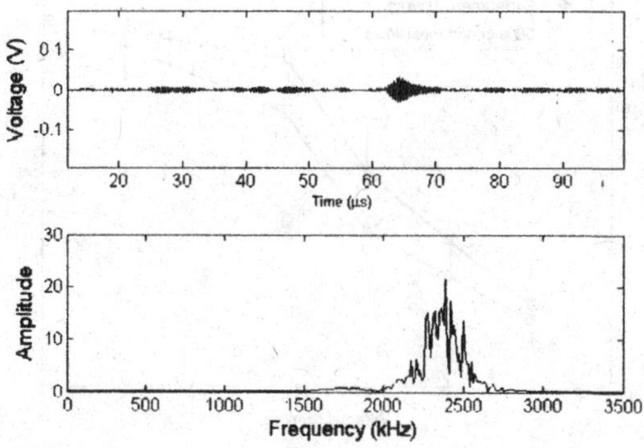

Figure 10. Signal shown in Figure 7 after being filtered with a pass-band filter between 2.2 and 2.6 MHz.

specimens with different level of adhesion, i.e., different pummel ratings, see Tables I and II. Based upon the results shown in Figures 6 and 7, the S_1 mode was selected to estimate the adhesion level for the laminated samples using a frequency range of 2.1 to 2.6 MHz.

The general arrangement of the experimental setup consisted of a pulser-receiver, a pair of piezoelectric angle-beam transducers and adjustable wedges, a filter, one analog-to-digital board and waveform storage software, see Figure 1. Both the sending transducer and the receiving transducer were located on the same side of the laminated glass specimen. A holding fixture was used to maintain an adjustable fixed separation distance between the sending and receiving transducers, and to allow the transducers to be moved as one unit. The transducers are broadband unfocused sensors with central frequency of 2.25 MHz. A wedge angle of 70^0 was chosen to excite the S_1 mode.

Figure 8 shows a transducer sensitivity test, i.e., the system response when the two transducers are held face-to-face (i.e., the two transducers are held together with zero distance) and using a pass-band filter from 1.5 to 3.0 MHz. Figure 9 shows the time domain signal and its frequency content associated with the S_1 mode using a transducer separation distance of 200 mm and a band-pass filter between 1.5 MHz to 3.0 MHz. Figure 9 shows that the frequencies in the range of 2.1 MHz to 2.6 MHz have been severely attenuated. This result is fully consistent with the attenuation curves shown in Figure 6. To estimate the adhesion levels, the signal was then further filtered around the frequency of 2.4 MHz. Figure 10 shows the same signal shown in Figure 9 after being further filtered using a band-pass filter from 2.1 MHz to 2.6 MHz using MatLab.

Figure 11 shows the predicted and the experimentally measured attenuation for the S_1 mode for laminates with different pummel ratings. The mean was obtained by averaging ten experimental data points. The red bars show the 90% confident interval calculated using the student 't' test. The results show a relatively good agreement between predictions and the experimental measurements with a minimum error of 4.2% for specimens with a pummel number 8 and a maximum error of 28.5 % for specimens with a pummel number 5. Considering the difficulties associated with measuring attenuation reliably, such as beam spreading, inconsistent coupling, non-uniform distribution of the contact area, and non-uniform distribution of adhesion levels, this level of agreement is encouraging.

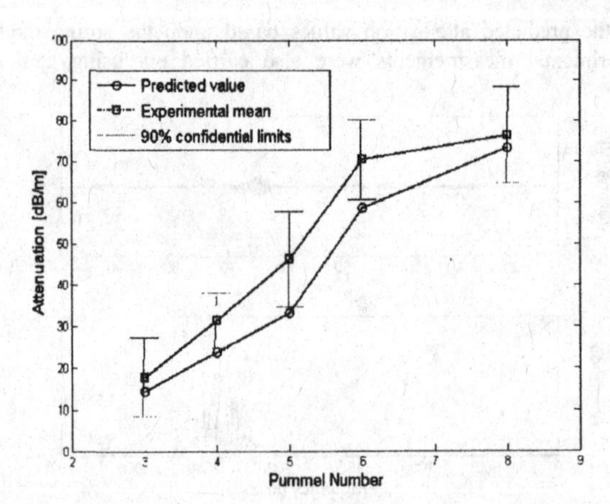

Figure 11. Predicted and experimentally measured attenuation for the mode S_1 versus pummel ratings for laminated safety glass test samples.

CONCLUSIONS

The imperfect interfaces between the plastic interlayer and the two adjacent glass plates in laminated safety glass were modeled using a bed of longitudinal and shear springs. The area of contact at the interfaces between the plastic interlayer and the two adjacent glass plates for each of the laminates was estimated via surface characterization experiments using atomic force microscopy measurements. Fracture mechanics concepts were then used to determine the stiffness values of the longitudinal and shear springs for laminates with different levels of adhesion. The attenuation of the S_1 mode for the laminates with different adhesion levels was then evaluated. The experimentally measured attenuation values of the S_1 mode at the frequency of 2.4 MHz showed relatively good agreement with the predicted values obtained from dispersion curve analysis. Laminates with higher adhesion levels were found to have corresponding higher attenuation values for the S_1 mode.

ACKNOWLEDGMENTS

This research was partially supported by the National Science Foundation under Grant No. CMS 06-23949. The support of Dr. Shih Chi Liu is greatly appreciated. The authors are grateful to Dr. John J. D'Errico from the Solutia, Inc. for manufacturing the safety glass specimens. The surface characterization experiments were carried out in the Materials Research Laboratory at the University of Illinois at Urbana-Champaign, which is partially supported by the U.S. Department of Energy under grant DEFG02-91-ER45439. The authors are also grateful to Mr. Scott McLauren for helping with the AFM surface measurements.

REFERENCES

[1]Y. Sha, C.Y. Hui, E. J. Kramer, P. D. Garrett, and J. W. Knapczyk, Analysis of Adhesion and Interface Debonding in Laminated Safety Glass, *J. Adhesion Sci. Technol*, **11**(1), 49-63 (1997).

[2]R. L. Eifert, and R. E. Moynihan, Fundamental Studies of the Adhesion Bond Strength of Butacite 106 to Glass, E.I. du Pont de Nemours and Co. Wilmington, Delaware, (1970).

[3]J. Svobota, and V. Matys, Polyvinyl Butyral Films for Laminated Safety Glass, *Plasty A. Kaucuk*, **12** (5), 131-135, (1975).

[4]*Monsanto Co.*, Polyvinyl Butyral Interlayer Laminating Guide, St Louis. MO, (1984).

[5]D. J. David, and T. N. Wittberg, ESCA Studies of Laminated Safety Glass and Correlations with Measured Adhesive Forces, *J. of Adhesion*, **17**, 231-244 (1984).

[6]J. R. Huntsberger, Adhesion of Plasticized Poly(vinyl Butyral) to Glass, *J. of Adhesion*, **13**, 107-129 (1981).

[7]J.E. Feig, and R. E. Moynihan, The Compressive Shear Test as a Measure of the Adhesion of 'Butacite' to Glass, E.I. du Pont de Nemours and Co. Inc. Wilmington, Delaware, (1971).

[8]MVSS 205, Motor Vehicle Safety Standard No. 205 Glazing Materials – Passenger Vehicles, Motorcycles, Trucks and Buses, Federal Safety Standards, (1977).

[9]ANSI Z26.1, Safety Code for Safety Glazing Materials for Glazing Motor Vehicles Operating on Land Highways. American National Standards Institute, Inc., (1977).

[10]S. Huo, and H. Reis, "Estimation of adhesive bond strength in laminated safety glass," Proceedings of SPIE Symposium and Smart Structures, San Diego, CA, 6529, 65290B-1 to 65290B-12, (2007).

[11]K. Balasubramaniam and C.A. Issa, Multi-Layered Interface Model for Simulation of Adhesive Bond Weakness Evaluation Using Ultrasonic Techniques, *ASME, Vibro-Acoustic Charact. of Materials and Structures*, **14** (1992).

[12]M. J. S. Lowe and P. Cawley, Applicability of plate wave techniques for the inspection of adhesive and diffusion bonded joints, *J. of Nondestructive Evaluation*, **13** (4), 185-200 (1994).

[13]A. Pilarski and J. L. Rose, Lamb wave mode selection concepts for interfacial weakness analysis, *J. of Nondestructive Evaluation*, **11** (34) 237-249 (1992).

[14]J. Baik and R. B. Thompson, Ultrasonic Scattering from Imperfect Interfaces: A Quasi-State Model, *J. of Nondestructive Evaluation*, **4**, 177-196 (1984).
[15]H. Tada, P. Paris, and G. Irwin, *The Stress Analysis of Cracks Handbook*, Del Research Co., St. Louis (1973).
[16]P. C. Paris, The Mechanics of Fracture Propagation and Solution to Fracture Arrester Problems, Document D2-2195, the Boeing Company (1957).
[17]B. Bhushan, *Introduction to Tribolog*, John Wiley & Sons, *New* York (2002).
[18]R. Seifried, L.J. Jacobs, and J. Qu, Propagation of Guided Waves in Adhesive Bonded Components, *NDT & E*, **35**, 317-328 (2002).
[19]P.C. Xu, and S. K. Datta, Guided Wave in a Bonded Plate: A Parametric Study, *J. Appl. Phys.*, **67** (11) 6779-6786 (1990).
[20]O. V. Rudenko, and C. A. Vu, Nonlinear Acoustic Properties of a Rough Surface Contact and Acousto Diagnostics of a Roughness Height Distribution, *Acoustic Physics*, **40** (4), 593-596. (1994).
[21]P.B. Nagy, Ultrasonic Detection of Kissing Bonds at Adhesive Interfaces, *J. of Adhesion Science and Technology*," **5** (8) 619-630 (1991).
[22]K. Kendall and D. Tabor, An Ultrasonic Study of the Area of Contact Between Stationary and Sliding Surface, *Proc. Roy. Soc. Lond.*, **323** (1554), 321-340 (1971).
[23]C.H. Yew and X. W. Weng, Using Ultrasonic SH Waves to Estimate the Quality of Adhesive Bonds in Plate Structures, *J. of the Acoust. Soc. of Am.*, **77** (5), 1813-1823 (1985).
[24]M. Lowe, "Matrix Techniques for Modeling Ultrasonic Waves in Multilayered Media." *IEEE Transactions on Ultrasonics, Ferroelectrics, and Frequency Control*," **42**, pp. 525-542, (1995).
[25]C. W. Chan and P. Cawley, "Lamb waves in highly attenuative plastic plates," *J. of the Acoust. Soc. of America*, **104**(2), pp. 874-881, (1998).
[26]F. Simonetti, "Lamb wave propagation in elastic plates coated with viscoelastic materials," *J. of the Acoust. Soc. of America*, **115**(5), pp. 2041-2053, (2004).
[27]L. M. Brekhovskikh, *Waves in Layered Media*, Academic Press, (1980).
[28]F. Simonetti and P. Cawley, "On the nature of shear horizontal wave propagation in elastic plates coated with viscoelastic materials," *Proceedings of the Royal Society of London, Series A*, **460**(2048), pp. 2197-2221, (2004).
[29]Bernard, M. J. S. Lowe, and M. Deschamps, Guided Wave Energy Velocity in Absorbing and Non-Absorbing Plates, *J. Acoust. Soc. Am.*, **110** (1), 186-196 (2001).
[30]B. Pavlakovic and M. J. S. Lowe, *Disperse User's Manual Version 2.0.11*, Imperial College, Univ. of London, (2001).
[31]J. L. Rose, *Ultrasonic waves in solid media*, Cambridge University Press, Cambridge, United Kingdom (1999).
[32]P. P. Delsanto, and M. Scalerandi, A Spring Model for the Simulation of the Propagation of Ultrasonic Pulses Through Imperfect Contact Interfaces, *J. Acoust. Soc. Am.*, **104** (5), 2584-2591 (1998).
[33]B. Drinkwater, R. D., Joyce, and P. Cawley, A Study of the Transmission of Ultrasound Across Solid-Rubber Interfaces, *J. Acoust. Soc. Am.* **101** (2), 970-981 (1997).

APPLYING MODELING TOOLS TO PREDICT PERFORMANCE OF TRANSPARENT CERAMIC LAMINATE ARMORS

C.G. Fountzoulas, J.M. Sands, G.A. Gilde, and P.J. Patel

U.S. Army Research Laboratory, Survivability Materials Branch, 4600 Deer Creek Loop, AMSRD-ARL-WM-MD, APG, MD, USA

ABSTRACT

The dominant materials technology used for ballistic transparency protection of armored tactical platforms in commercial and military applications is low cost glass. Currently, the development of next generation ceramics is a key to providing enhanced protection capability and extended service performance for future armored windows. A development program is underway to improve processing and reduce costs associated with transparent ceramic materials, such as magnesium aluminate spinel, $MgAl_2O_4$. A critical element of this development is to understand the influence of defects on the failure performances, both statically and dynamically. Although commercial efforts continue to focus on critical cost reduction and processing parameters, interest continues to grow toward applying ceramics toward armor system design. Future ballistic designs are being assembled in a virtual environment to begin to extrapolate the importance of defects, including flaw density, crystal orientation, shape and size, position with respect to the ceramic surface, and ceramic plate dimensions on the performance of ceramic spinel under dynamic impacts. While additional developments remain to be completed, the successful development of process methods that result in consistent high quality transparent ceramics is allowing validation of models to increase the insertion potential for these transparent ceramics into next generation tactical military platforms. The purpose of this report is to demonstrate the importance of modeling tools in advancing ceramic transparent armor materials to fielded applications. The effect of various shape defects, located in the interior and on the surface of spinel, on the failure of the transparent material will be discussed in detail.

Keywords: armor, modeling, spinel, failure analysis, defects.

INTRODUCTION

Transparent armor systems using ceramics as the striking face have been explored since the early 1970's because they potentially provide superior ballistic protection to conventional glass based transparent armor systems [1]. However, commercial manufacturers have not experienced a demand for large transparent (>1290 cm^2) ceramic plates for commercial markets in point-of-sale scanners and fluorescent lighting. Therefore, the development of large (> 645 cm^2) and thick (>12.7 cm) ceramic plates for transparent armor applications has not proceeded to commercial maturity. The U.S. Army has invested heavily in the development of next generation materials, including ceramics, for military systems [2]. The result of the on-going investments is a critical understanding of ceramics strengths and weaknesses for military platforms.

As large transparent ceramic materials are available commercially in sizes up to 900 cm^2, progress in ballistic designs has offered substantial increases in performance in transparent armor design. Among the potential ceramic materials considered for armor — sapphire, edge-form-growth sapphire, magnesium aluminate spinel, aluminium oxynitride — one was selected for the current pursuit, magnesium aluminate spinel ($MgAl_2O_4$). Although individually, single impact testing has shown small variations in ballistic efficiencies (<10%) of ceramics, to achieve multi-hit performance, all of the ceramics are effectively equivalent. Technology Assessment and Transfer (Annapolis MD,

USA) is providing ceramic spinel plates produced via hot-pressing in sizes up to 28 cm x 36 cm x 1.5 cm for this report [3].

Finite element modeling has progressed substantially in the ability to predict failure of materials under extreme dynamic loading conditions. One of the limitations of predictive models is lack of a complete dynamic materials properties database which is needed for materials models for each of the materials in the simulations. In order to compensate for parameters whose dynamic values were extrapolated from their static or quasi-static properties, baseline experiments are often used to recalibrate the models. [4, 5] However, the recalibration method of modeling lacks many of the physical properties and failure mechanisms associated with real-world materials. Therefore, often recalibrated models lack the ability to predict within statistical error future failures over any substantial ranges due to the existence of defects, and materials substitutions often lead to new calibration requirements. The desired approach is to validate a fully characterized materials database with one calibration model, and subsequently apply the model to modified designs. However, despite its apparent problems, recalibration of existing materials models has been proven to be an effective tool in the hands of the modeler by minimizing the number of simulation iterations and resulting in more successful predictions. Regardless of methodology, finite element tools can be applied effectively to reduce the variability between impact tests and can be used to validate designs with fewer experimental failures, when robust models are created. [5]

EXPERIMENTAL

Due to the sensitive nature of ballistic test results, surrogate materials were assembled that do not represent current or future armored designs.

Ballistic Coupons

Experimental coupons for ballistic testing consisted of laminated layers of spinel bonded using Huntsman 399 polyurethane adhesive to Bayer polycarbonate. The laminate sandwich was assembled in an autoclave at 93°C to 121°C for four hours. To reduce the variables in the investigation, the backing layer thickness was fixed at 1.27 cm of polycarbonate. The ceramic striking material is varied between 1.1 cm and 1.5 cm thick, depending on the density of flaws introduced. The bonding layer is typically 0.10 cm. Experimental samples were evaluated only to attain penetration velocity to confirm the model parameters. In addition, square cuts of 1.5 mm width and 5 mm height, and cones of 4 mm diameter and 4 mm height were introduced into the surface of the spinel. The density of the surface defects varied and it represented a 2% and 4% mass loss of the solid spinel. Otherwise, all results and discussion, including the internal defects architecture, are based on simulation analysis.

Modeling

The ballistic behavior of an identical to the actual target geometry, which was consisted of spinel, polyurethane and polycarbonate, and impacted by a surrogate projectile, was simulated using the non-linear ANSYS/AUTODYN commercial package [7]. The material models used were obtained from the AUTODYN library. The modeling laminated target consisted of panels of spinel, polyurethane, and polycarbonate of 900 cm^2 cross sectional area (30 cm in 2D models). The defects were filled with air at one atmospheric pressure. Some of the strength material models of the laminated target materials had been previously recalibrated using existing ballistic data. The numerical modeling of undamaged armor coupons, and the coupons with surface defects, was carried out in two and three dimensions. The projectile applied in the models was a 3.0 cm long, 1095 steel projectile, of conical frustum geometry, (6 mm large base, 1.0 mm small base), using two dimensional axi-symmetric models and plane symmetry models. Depending on the geometric complexity of the laminated target and the existing pre-processing capability of the solver, smooth particle hydrodymamics (SPH), Lagrange and Euler solvers were used. In particular, when the SPH solver was used the target and the projectile were discretized by smooth particle hydrodynamics (SPH) with a particle size of 0.2 mm for increased modeling accuracy. The element size used for the 2D Lagrange and Euler solver was 0.25 mm. Since each of those solvers has its own characteristics, to ensure result compatibility a target containing no defects was simulated by all three solvers. All solvers produced similar results. Results were obtained by simulating projectiles impacting the targets at the experimental velocities ranging from 500 to 1000 m/s.

Defect Models

One of the advantages of modeling methods is the ability to create physically challenging architectures to investigate effects of point defects on failure. The sensitivity of ballistic measurement tools is typically less than ±10% due to the range of available failure modes invoked in the high energy exchange between projectile and target. Additionally, capturing the real-time failure modes in the impact event requires highly specialized video equipments. These factors contribute to a very difficult and expensive set of experiments for investigating small flaws and the impact on performance in the experimental realm. By using modeling tools, however, the effects of macroscopic flaws and the location of these flaws can easily be investigated in a model that correctly captures the physics of failure in the materials. Therefore, to enhance the understanding of flaws and the behavior of spinel with defects, the modeling approach, which had been validated previously by experimental data for the case of surface introduced discontinuities, defined also as defects, is employed. Figures 1-2 show internal defects of various cross-sectional area introduced in the spinel.

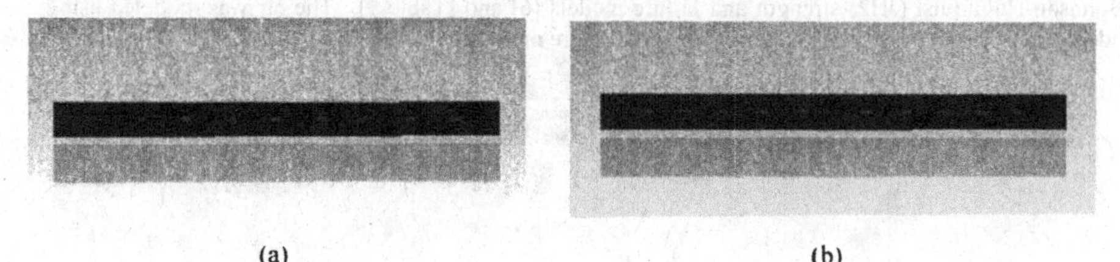

(a) (b)

Figure 1. Target containing (a) 9 internal elliptical defects; and (b) 10 internal elliptical defects.

Consideration of defects in calculating the failure probability of ceramic articles with short-term loading has been reported by Gorbatsevich et al [7]. The effect of various shapes and locations of defects of equal cross sectional area was studied by simulation of the impact. The location of the defects at either surface or interior to the sample is considered, and defects shapes from elliptical, to square, to circular are explored. Due to limitations of the SPH solver preprocessor, for this study the Lagrange and Euler solvers were used. However, the Lagrange solver was not able to be used for all shapes, due to its known sensitivity to the erosion of cells of shapes with sharp corners, like the elliptical and rectangular cross sectional areas defects. The 2-D Euler solver produced results for all defect shapes. To ensure compatibility of our previous results produced by the SPH solver and between the Lagrange and Euler solvers as well, we compared the results of the impact simulation for a target without defects which were run under identical impact and boundary conditions. All three solvers produced similar results. In the interest of computational consistency, defect results were compared using residual velocities based on a constant initial impact velocity. To keep simulation times reasonable, all specimens are impacted at an initial velocity of 975 m/s.

The effect of surface flaws on the ballistic performance of the target was studied by introducing square cuts into the surface of the spinel of 1.5 mm width and 5 mm height and comparing the results against a target without defects. The effect of the defects on areal density of the coupons is less than

1%. Simulation images after impact are shown in Figure 3. The "defects" were investigated both at the exposed surface and positioned on the interlayer (or directly on PC when the bonding layer is excluded). Five and nine "defects" were introduced on the spinel, corresponding to 2% and 4% mass loss respectively. The equal area interior defects to the surface defects were introduced along the geometric centerline of the spinel target spaced equally from each other. The mass loss of the internal defects corresponds to 4%. The defect closer to the outer edge of the spinel was placed at a distance equal to the surface defect from the outer edge of the target. The projectile impacted the target in the following ways: (a) perpendicular to the center of middle defect; and (b) perpendicular to the target but offset from a defect, midway between two defects. Analysis of the experimental tests and the simulations included evaluation of the extent of the damage, residual velocity after impact, and extent of deflection. 3D simulations of a laminated target with and without surface defects were also carried out.

The material models used for the polycarbonate, urethane interlayer, and steel projectile were obtained from the AUTODYN material library [6]. The PC was modeled using a shock equation of state (EOS), piecewise Johnson-Cook (JC) strength model, and a plastic strain failure criterion. The urethane was modeled using a linear EOS and a principle stress failure criterion. The projectile steel was modeled using a shock EOS and a JC strength model [6]. The spinel, however, due to lack of existing material model, was modeled using a recalibrated form of the existing AUTODYN library Al_2O_3 material model, produced from existing experimental data and consequently validated many times by predicting within 3% results of new target design, which included a polynomial EOS and Johnson-Holmquist (JH2) strength and failure models [6] and [Table 1]. The air was modeled using ideal gas equation of state, with no strength and failure models.

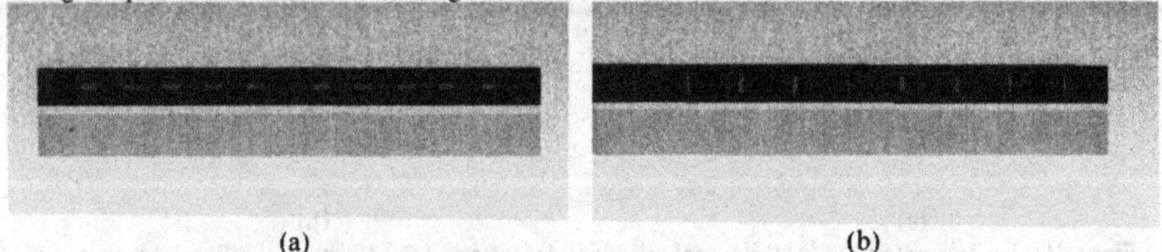

(a) (b)

Figure 2. Target containing (a) 10 internal rectangular defects; and (b) 9 rectangular internal defects with the larger dimension of the cross-section parallel to the line of impact.

RESULTS

Surface Defects

The minimum penetration impact velocity for the steel projectile into the baseline, defect free spinel/polycarbonate target was 850 m/s. The result is validated using the experimental target, with excellent agreement. It is noted that the simulation results were completed more than 30 days prior to experimental results and agreement was within 3%. Therefore, the failure criteria and confidence in the material parameters is excellent. The extent of damage in the simulation result does not coincide precisely with the experimental coupons for this case. The simulation results show potential edge effects that do not appear in the experimental system. It should be noted, however, that during the impact event, failure modes appear consistent. Therefore, the 2D model shows an effective and rapid method of producing laminate constructions for interrogation of failure criteria. The baseline results allow confident investigation of the defect models. That prediction disagreement may be attributed to the fact that a 2D defects, when expanded in 3D is not a localized 3D defect, but rather a groove. Therefore, more material is removed at a 2D simulation than the 3D respective simulation. Currently performed 3D simulation on targets containing conical surface defects tend to confirm this hypothesis.

Moreover, the simulation almost duplicated the experimental surface cracking. The authors believe that a very accurate reproduction of a surface cracking is possible by using a SPH particle size smaller than the currently used of 1 mm.

Strength	Johnson-Holmquist
Shear Modulus	1.35000E+08 (kPa)
Hugoniot Elastic Limit	5.90000E+06 (kPa)
Intact Strength Constant A	9.89000E-01
Intact Strength Exponent N	3.75500E-01
Strain Rate Constant C	0.00000E+00
Fractured Strength Constant B	7.70000E-01
Fractured Strength Exponent M	0.90000E+00
Max. Fracture Strength Ratio	6.00000E-01
Failure	**Johnson Holmquist**
Hydro Tensile Limit	-2.90000E+04 (kPa)
Model Type	Continuous (JH2)
Damage Constant, D1	1.00000E-02
Damage Constant, D2	1.00000E+00
Bulking Constant, Beta	1.00000E+00

Table 1. Strength and failure parameters of spinel (modified Al_2O_3 from the AUTODYN library)

When investigating the effects of the urethane interlayer, the overall velocity changes in the baseline case are insignificant. The interlayer material does not add sufficiently to the mass of the system to impact performance metrics established. Further, the thickness of the system, and the selected thickness of the interlayer do not provide significant attenuation of any blast or shock waves in the system. Therefore, the defect cases of the simulation are run with the interlayer present to adequately represent the thickness and stiffness effects of the experimental coupons.

When the defect targets were evaluated, the performance was dramatically lower.. The overall effect of performance has yet to be fully interrogated. However, in principal, a notched or defect based target should allow localized failure in the ceramic strike face and, therefore, limit the extent of damage in neighboring cells. On initial investigation, this conceptual hypothesis does not appear to hold true. However, the projectile receives a much lower interaction potential with the ceramic and is less eroded during the initial phases of the impact event. The resulting exit velocities of the projectile are dramatic. For a baseline target with no penetration, inclusion of 5 defects at the urethane surface, impacted by a projectile of initial velocity of 1000 m/s, produced an exit velocity of 70 m/s. This is a 7% reduction in efficiency. Further, the damage area, as estimated by length of damage in the sample, appears to be comparable or larger with the defects. The effects are more dramatic as the number of defects is increased. The simulations showed that addition of 5 defects and 9 defects results in a residual velocity of 95 and 150 m/s or a 9.5% and 20% reduction in efficiency respectively. Figure 3a and 3b show damage extent on impact of 5 and 9 defect samples at 11 microseconds after impact.

(a) (b)

Figure 3. Snapshot images of simulations of damage propagation into spinel-urethane-polycarbonate stacks after 11 μs with slug moving 800 m/s with a) 5 notches and b) 9 notches.

The experimental evaluation and numerical modeling of the impact of a laminated target with a spinel strike face and a polycarbonate backing showed that the resistance of the target to penetration deteriorates with the presence of surface defects. The position of these defects relative to a projectile did not appear to dramatically effect the performance reduction. The deterioration of the target seems to depend somewhat on the amount of defect in the target. which is related to the mass removed and to the properties like system stiffness, overall density and residual stress from the lamination process. However, the modeling also showed that the extent of the damage is only partially confined by the inclusion of intentional defects. The two dimensional slits employed in this analysis correspond to grooves in a three dimensional target. Since the target integrity depends on the material removal we anticipate that the 3D simulation of holes of same cross sectional area of the grooves will result in lower exit velocity and potentially different damage patterns. It is worth examining the effect of the defect orientation and size on the ballistic performance of the target. The simulations that were performed with PU interlayer indicate that the presence of the polyurethane provides an additional resistance to penetration for the samples with defects. but the overall effect did not alter the results in a measurable way.

In the case of the spinel surface consisted of conical defects the position of these defects relative to the line of impact produced different results. 2D and 3D simulations showed that for the case of the projectile impacting directly a cone. the exit velocity was close to the exit velocity from a spinel with no defect. However. when the projectile impacted the spinel between defects then the exit velocity was close to the exit velocity from a spinel containing defects at the side next to the polyurethane.

Internal defects

The simulation showed clearly that the internal defects have a reduced effect on the target resistance to penetration when compared to surface defects. The overall damage of the spinel at 100 μs appeared to be similar for all defect shapes (Figure 4a and 4b). Simulations also showed that internal defects of elliptical cross section seem to decrease the penetration resistance of the target more when compared to the circular and rectangular penetration. In addition, the simulation showed that a defect of elliptical cross section with its large axis parallel to the line of impact resulted in the largest exit velocity. thus smaller resistance to penetration. Our results indicate that the larger decrease of the resistance to penetration caused by the presence of the surface defects may be attributed to the damage wave traveling through intact spinel before its exit from the spinel to the polyurethane. It also appears that the spinel deterioration during impact depends heavily on the material removal caused by the presence of voids. The defects under or very close to the impactor became deformed during the impact and finally they collapsed.

The damage progression seemed to grow continuously until the midpoint of two neighboring defects. However. at this point the damage appeared at the next defect while it continued to grow.

This phenomenon was more obvious at defects with sharp corners, such as elliptical and rectangular cross section (Figures 5a to 5b). This may be attributed to the damage wave arriving at the sharp corners, points of stress concentration, faster than the progressing damage. The damage after it encountered the first defect by-passed initially the next defect and continued from the third defect (Figure 8). Finally, the spinel target deteriorated as the impactor continued to travel through it.

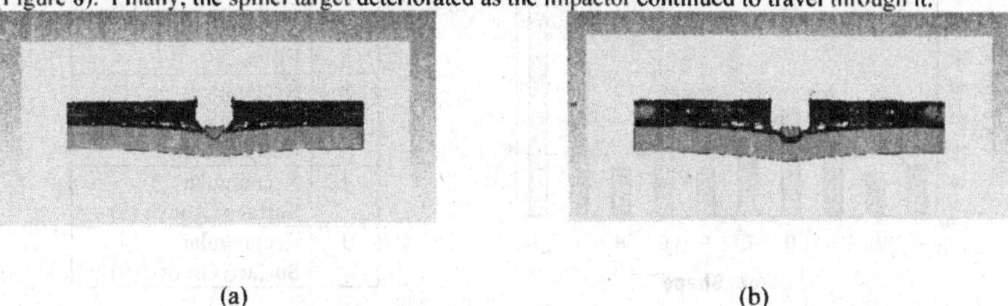

(a) (b)

Figure 4. a) Damage of spinel containing 9 circular defects after 100 μs; (b) damage of spinel containing 10 rectangular defects after 100 μs. The damage extent for both cases appears to be similar.

Figure 8 shows the effect of the various-shape, equal-area, defects on the exit velocity of a projectile impacting a target with a velocity of 975 m/s. The simulations showed that the surface defects decrease the resistance to penetration more than the internal defects. The presence of surface defects resulted in larger exit velocity even when compared to internal defects which correspond to double mass loss [Fig. 6, compare Surface Groove 5 (2% mass loss) to any of the internal defects (4% mass loss)].

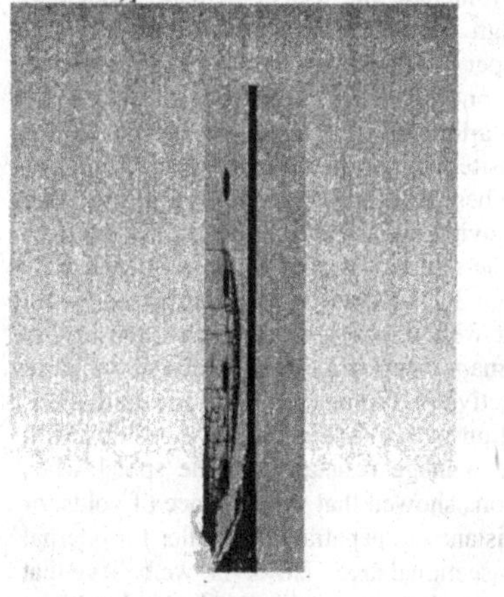

(a) (b)

Figure 5. (a) Damage progression after 30.37 μs; (b) damage progression after 32 μs. These two slides show that the damage appeared at the sharp edges of the next defect as soon as the damage reaches the midpoint between the two defects.

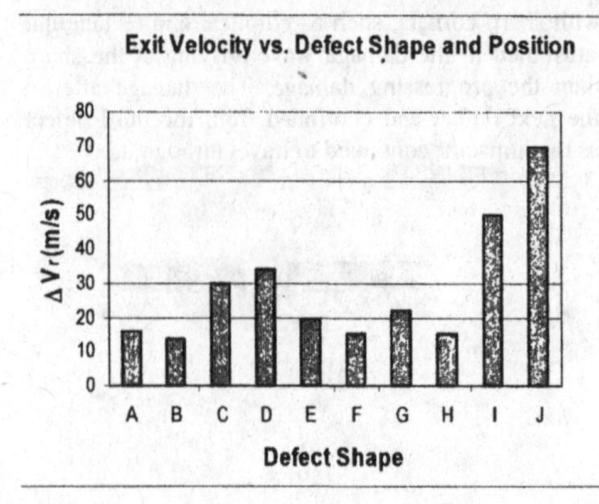

A	Circle 9
B	Circle 10
C	Ellipse 9
D	Ellipse 9r
E	Ellipse 10
F	Rectangle 9
G	Rectangle 9r
H	Rectangle 10
I	Rectangular Surface Groove (5)
J	Rectangular Surface Groove (9)

Figure 6. Residual velocity for spinel/polycarbonate targets impacted with a steel penetrator at 975 m/s. where defects are purposefully included in the ceramic layer with specific design parameters keeping volume of defects constant: and compare effect of internal defects to surface defects.

CONCLUSIONS

While the need for advanced materials solutions for protection of vehicles from ballistic threats continues to grow. the ability to predict material performance using advanced modeling tools increases. The current efforts underway in the U.S. Army include the use of ballistic modeling, ballistic testing. and historic knowledge of ballistic design to create structural armors for the transparent armor needs of the U.S. military. The current paper has demonstrated the powerful use of computational modeling to predict the effects of defects on failure in ceramic materials. The increasing density of flaws, simulated as cuts or slits of various equal area shapes, resulted in a significant reduction in apparent local stiffness in the composite laminates, and resulted in significant changes in the virtual performance of the laminate stacks. These results were verified using ballistic testing to demonstrate the complex nature of the ballistic environment. While the surface grooves represent only a 2-4 % reduction in the materials content in the samples. they resulted in 7% and 20% reduction in performance, and compare this with the maximum 3.5% reduction in performance caused by the elliptical internal defect (Fig. 8). This is consistent with previous research that shows that ceramics with flaws demonstrate reduced mechanical performance during static testing. and validates the potential use of random flaws in model systems for predictive performance in dynamic validations using simulation. The presence of surface flaws in conjunction with the defect relative position with respect to the line of impact was more detrimental to the damage resistance of the spinel when compared to the internal of equal area defects. The simulations showed that the presence of voids. or the mass decrease caused by defects. results in decreased resistance to penetration, smaller for internal defects and relatively larger for surface defects of equal cross-sectional area. However, we believe that as the defect density increases the resistance to penetration and the ceramic deterioration will be accelerated. It is worth studying the effect of defect clustering with respect to the line of impact, defect density, defect shape irregularity, orientation and inclusions of various nature materials rather only air.

ACKNOWLEDGEMENT

The authors wish to thank the reviewers wholeheartedly for their successful suggestions and corrections which made this paper better.

REFERENCES

[1] Gatti,A & Noone, M J, Feasibility Study for Producing Transparent Spinel (MgAl2O4). AMMRC-CR-70-8, February 1970.
[2] 2006 Army Modernization Plan. "Building, Equipping, and Supporting the Modular Force," Annex D. March 2006.
[3] http://www.techassess.com/tech/spinel/spinel_prop.htm, 20 September 2007.
[4] C.G.Fountzoulas. B.A. Chéeseman, P.G.Dehmer and J.M.Sands, "A Computational Study of Laminate Transparent Armor Impacted by FSP", Proceedings of 23rd Inter, Ballistic Symp., Tarragona, Spain, 14-19 April 2007
[5] C. G. Fountzoulas, J.C.LaSalvia, B.A.Cheeseman, "Simulation of Ballistic Impact of a Tungsten Carbide Sphere on a Confined Silicon Carbide Target", Proceedings of 23rd Inter, Ballistic Symp.. Tarragona, Spain, 14-19 April 2007
[6] ANSYS/AUTODYN Vol 11.0, Manual, Century Dynamics Inc., Concord, CA
[7] M. I. Gorbatsevich and A. E. Ginzburg, Strength of Materials, pp. 392-397. 23 (4), April, 1991

AN ECONOMIC COMPARISON OF HOT PRESSING VS. PRESSURELESS SINTERING FOR TRANSPARENT SPINEL ARMOR

A. LaRoche, K. Rozenburg, J. Voyles, L. Fehrenbacher
Technology Assessment & Transfer, Inc.
133 Defense Highway, Suite 212
Annapolis, MD 21401

Gary Gilde
US Army Research Laboratory
AMSRL-WM-WC Building 4600
Aberdeen, MD 21005

ABSTRACT

The study makes an economic comparison between two competing processes for the manufacture of large spinel plates suitable for transparent armor. The study is based on actual and estimated cost data applied to the state of the art processes practiced by Technology Assessment & Transfer (TA&T) to manufacture transparent spinel components. Cost-related factors under consideration include materials, labor, product yields, and amortization of capital equipment. A most significant difference between the two processes is that in hot pressing, all of the sintering shrinkage occurs through the thickness and under conditions of plasticity; whereas in sintering, the shrinkage occurs in three dimensions and much of it under non-plastic conditions. Hot pressing inherently has higher product yields due to reduced cracking and warping of the plates. Hot pressing is TA&T's preferred manufacturing method for large, transparent spinel armor windows for economic reasons and also for overarching reasons of feasibility when very large windows are considered.

OVERVIEW OF SPINEL TRANSPARENT ARMOR

For all threats greater than handguns, transparent armor systems are based on transparent ceramic materials, of which the oldest and most common is glass. Newer transparent ceramics include polycrystalline materials such as magnesium aluminate spinel (spinel) and aluminum oxynitride (AlON). Single crystal aluminum oxide (sapphire) is also a viable transparent armor material. All of these crystalline ceramics are significantly harder and stronger than glass, which makes them much more efficient as armor. Higher efficiency allows the armor system to be thinner and lighter weight, which in many cases, could be enabling factors for current and future ground and air platforms.

The fundamental requirements for transparent armor systems are transparency and resistance to ballistic threats. The major performance goal of transparent ceramic armor is the capability to defeat multiple ballistic impacts with minimum distortion of surrounding areas. Spinel has the lowest density of these three hard transparent ceramic materials, which gives it an advantage in mass efficiency and weight, as the ballistic performance of all three materials is essentially equivalent. Photographs of a transparent armor system built with a TA&T transparent spinel strike face and its multi-hit capability against a 7.62 mm armor piercing incendiary threat is shown in Figure 1 and Figure 2, respectively.

Figure 1. Transparent armor system using spinel strike face.

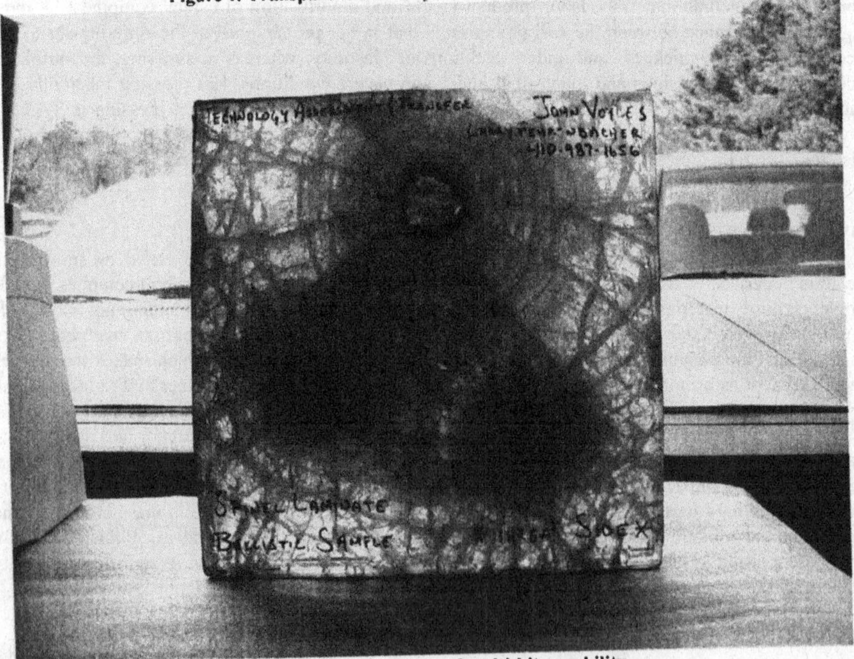

Figure 2. Demonstration of multi-hit capability.

Two other important factors are size capability and cost. Many ground platforms require large windows for mission capability; vehicles can be fielded with 15" x 20" and 16" x 40" windows. Providing these large windows as armor systems will be a challenge for any supplier of crystalline ceramics. Cost is always critical, and cost is the primary subject of this presentation.

THE MANUFACTURING PROCESS COMPARISON

Spinel raw material is readily available from a commercial supplier, with more suppliers coming online. In response to a challenge from Army Research Laboratory, Technology Assessment & Transfer delivered 20 hot pressed transparent spinel plates, each 14" x 14". This was accomplished in TA&T's pilot plant facility located in Millersville, MD. These armor plates were manufactured using the company's 250-ton hot press, but TA&T also has extensive experience manufacturing transparent spinel components via a traditional pressureless sintering (PS) process. TA&T recently conducted an analysis to compare hot pressing and pressureless sintering as competing methods for manufacturing transparent spinel armor plates. The analysis was primarily based on economic considerations, but feasibility was also considered where appropriate. Because of the experience recently acquired in the armor challenge, 14" x 14" plates were used as the basis for comparison. For purposes of comparing capital equipment requirements, a production rate of 10,000 plates per year was assumed. As the first step in the analysis, a side-by-side flow chart was constructed for the two competing processes, as shown in Figure 3.

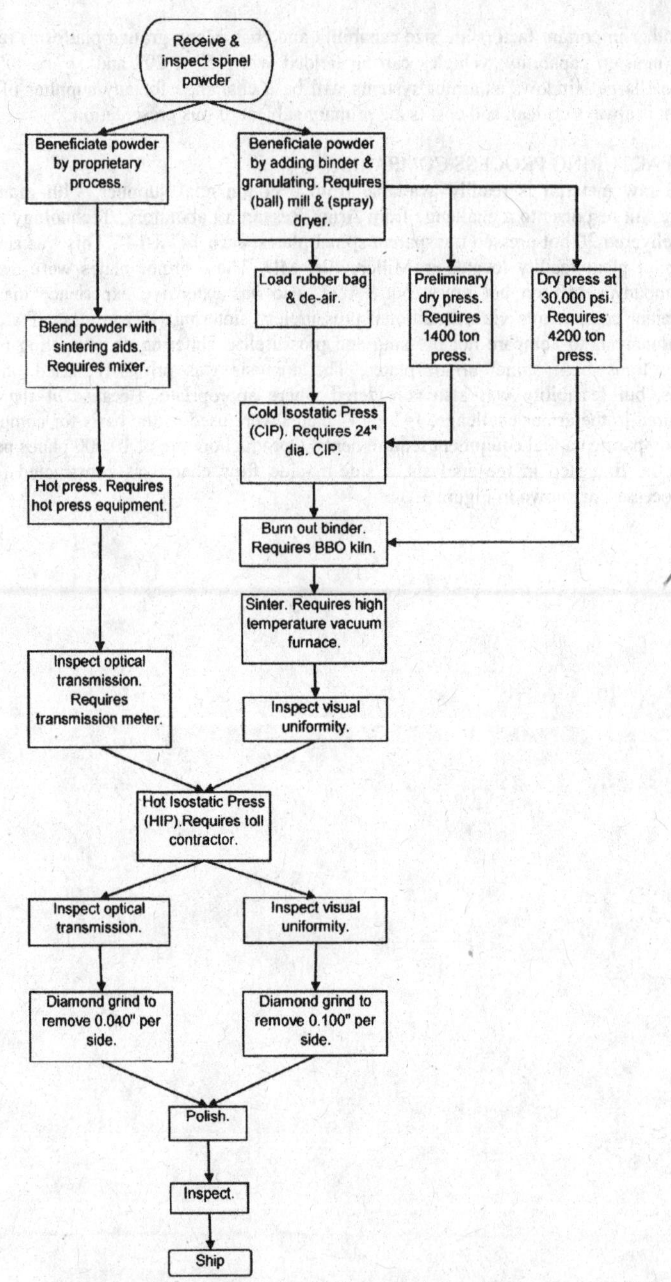

Figure 3. Side-by-side flow chart of hot pressing vs. pressureless sintering.

Each operation listed on the flow chart was analyzed, and the step-by-step comparisons are as follows:

The raw material currently used in both processes is S30-CR, a high purity spinel powder manufactured by Baikowski International. A common inspection procedure is used regardless of the intended manufacturing process.

Both processes require that the spinel powder be beneficiated, but in different ways. For hot pressing the powder is subjected to a proprietary beneficiation process for which the capital investment is about $250,000. Typically, the spinel powder for hot pressing is blended with sintering aids using a mixer costing $25,000 or less. For pressureless sintering a pressing binder is added and the powder is granulated by spray drying. The necessary ball mills and spray dryer would cost approximately a million dollars. The capital and labor costs associated with the additional spray drying step required for PS give hot pressing an additional cost advantage over pressureless sintering. It has also been demonstrated that good results can be achieved by hot pressing a lower cost spinel powder with lower surface area.

TA&T's 250-ton hot press is capable of producing 14" x 14" plates. Replacement cost for this hot press is estimated to be $500,000. TA&T has demonstrated that the PS process for manufacturing transparent spinel requires that the powder must be compacted at approximately 30,000 psi. At 20% shrinkage (an optimistic estimate), a 14" x 14" finished plate must be green formed to 16.8" x 16.8". Dry pressing a green part in a die would require a press rated at over 4200 tons. Such presses are rare and costly, on the order of $3 million. Obviously, cold isostatic pressing (CIPing) may be a better way to apply such high pressure to form the powder into the desired shape.

One method to CIP a plate is to perform a preliminary dry pressing operation at about 10,000 psi, seal the under-pressed plate into a bag, and then CIP it at full pressure. This approach would require a 1400 ton press, which is still relatively costly and hard to find.

More commonly, a rubber bag can be filled with powder and CIPed. This operation is relatively labor intensive, as the bag must be assembled into a support frame, filled, sealed and de-aired manually. It has also been observed that such a CIP procedure always produces a plate with significant variation in thickness. Generally, the center of the plate is thicker than the edges. This thickness variation requires that significantly more grinding stock be provided on the faces of the plate. A 24" diameter x 60" deep CIP could produce about 6 plates per cycle and would cost approximately $750,000.

The formed green part for the PS process must undergo a binder burnout step prior to sintering. Because the spinel powder has such high surface area and small particle size, binder removal is difficult and requires a long, slow burnout, perhaps as long as a week. A $10,000 ring furnace would burn out multiple plates at a time.

The PS plate must be sintered in vacuum. As noted previously, the plate undergoes at least 20% shrinkage in every dimension during sintering. This causes the edges to move more than an inch laterally on a 14" finished dimension. Such a large shrinkage can cause cracks and warpage, resulting in product loss. Hot pressing is also done in vacuum. Sintering forces are assisted by externally applied pressure, and by sintering aids that have proved effective only in hot pressing. During hot pressing all the shrinkage occurs through the thickness. The faces of as-hot-pressed plates are very flat and parallel compared to PS plates formed by CIPing. Hot pressing of multiple plates is accomplished by stacking the plates with spacers between.

Currently, all transparent spinel armor plates are hot isopressed (HIPed) to improve optical properties. TA&T contracts the HIP operation to Bodycote, and the service is quite costly. Many hot pressed plates come close to acceptable transparency without HIPing, and it is possible that process improvements will eventually improve as-hot pressed transparency to the point that HIPing will no longer be necessary. In the meantime, as-hot pressed plates can be inspected using legitimate optical properties such as light transmission. This allows early rejection of the rare plate that has poor quality.

PS plates are opaque, or translucent at best, after sintering. It is unlikely that the HIP step could ever be eliminated. The inspection of as-sintered plates is limited to gross physical defects, and optical transmission measurements are meaningless. Therefore, the PS process will potentially have lower yields after the costly HIP operation.

The post-HIP inspection is also more effective for hot pressed plates. The faces are relatively smooth and uniform, allowing the application of refractive index matching fluid and transparent covers, which, in turn, allow meaningful measurements of optical properties with a Gardner HazeGard®. The surfaces of PS plates are not flat or smooth enough to allow use of this inspection technique.

All transparent spinel plates must be diamond ground and polished to achieve full transparency. These processes are nominally the same for hot pressed and PS plates; however, the hot pressed plates require much less grinding because the faces are flatter and more parallel. The difference in grinding costs is estimated to be approximately $400 per plate. As described earlier, hot pressed plates will have a higher yield from the grinding and polishing operations because of the more effective post-HIP inspection.

For the postulated production schedule of 10,000 plates per year, each sized at 14" x 14", the hot press process is preferred. The estimated savings of $400 per plate in grinding costs alone amounts to a difference of $4 million in a single year. Labor costs are also less for hot pressing because it is a simpler forming process with fewer steps. Simpler and fewer process steps coupled with effective go-no go intermediate inspections will reduce rejections later in the spinel production cycle and thereby reduce finished armor costs. Both processes require significant investment in capital equipment. Large presses and large kilns are required, either separately or combined into hot presses.

Hot pressing is also preferred on the basis of process scale-up capability. It is obvious that lateral shrinkage is an overarching concern for large pressureless-sintered plates. For the 16" x 40" windows desired for the HEMTT vehicle, the as-formed plate would need to be 19" x 48", and each end of such a plate must move 4" during sintering. The challenges of making such plates flat and of uniform thickness are expected to be insurmountable. Conversely, hot pressed plates experience no lateral shrinkage, and the faces are as flat and parallel as the punches that form the plates. Plate yields are expected to be much higher for hot pressing, and grinding costs will be lower for all large sizes.

Figures 4 and 5. TA&T hot pressed spinel plates.

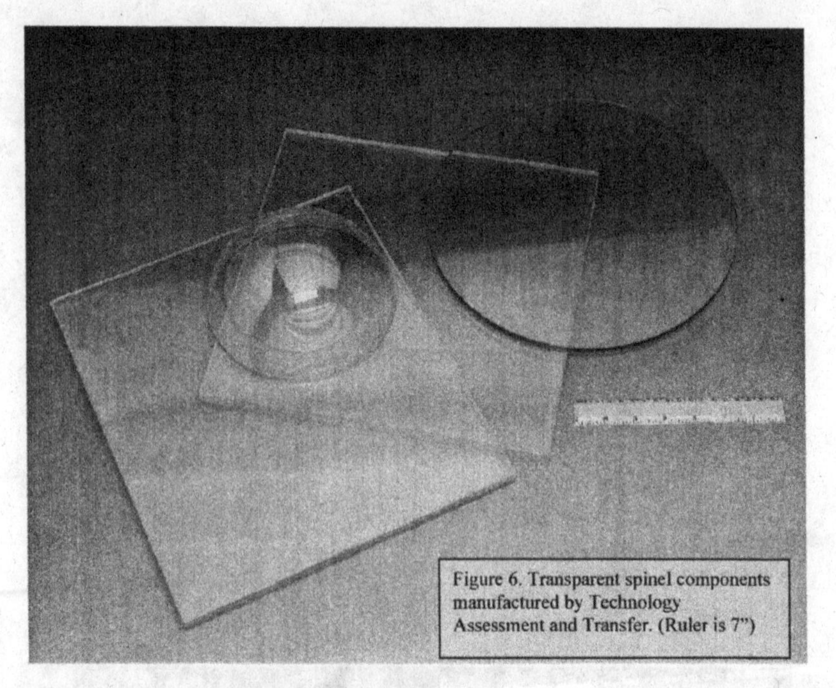

Figure 6. Transparent spinel components manufactured by Technology Assessment and Transfer. (Ruler is 7")

SUMMARY

TA&T has chosen hot pressing as the surest and most economical process for manufacturing large transparent spinel plates. The cost study clearly indicates that hot pressing offers the best chance of achieving the lowest cost per square inch for transparent crystalline ceramic strike faces. Although the 14" x 14" size may well be achieved by both processes under consideration, hot pressing will be much more feasible than pressureless sintering for the larger sizes that are anticipated in the future.

ADVANCES IN BALLISTIC PERFORMANCE OF COMMERCIALLY AVAILABLE SAINT-GOBAIN SAPPHIRE TRANSPARENT ARMOR COMPOSITES

Christopher D. Jones, Jeffrey B. Rioux, and John W. Locher,
Saint-Gobain Crystals
33 Powers Street
Milford, NH, 03055

Vincent Pluen, and Matthias Mandelartz
Saint-Gobain Sully
16 route d'Isdes
Sully sur Loire, France

ABSTRACT

Saint-Gobain has developed a ceramic transparent armor composite using our Edge-defined Film-fed Growth (EFG™) Sapphire crystals to defeat various ballistic projectiles. The high strength, high hardness, and good transparency in the visible and infrared spectra make sapphire an ideal choice for transparent armor applications. The composites are commercially available in large sizes (greater than 300 mm wide and 775 mm long) with substantial weight and thickness improvements over classical laminated glass solutions. This paper will discuss the properties and benefits of using Saint-Gobain sapphire transparent armor composites, along with the ballistic results, composition thickness, and areal density for defeating single shot and multi-hit ballistic results of armor piercing rounds (up to and including STANAG Level IV), and Fragment Simulated Projectiles.

INTRODUCTION

Saint-Gobain is the sole supplier of integrated ceramic transparent armor in the world.[1,2] The combination of expertise in Saint-Gobain Crystals (single crystal and ceramic materials) and Saint-Gobain Sully (lamination and ballistics) has resulted in products pushing the limits of ballistic performance, thickness, and weight. Saint-Gobain Crystals has been producing large sapphire sheets since 2003, growing thousands of crystals proving the commercial viability of large transparent ceramic materials.[3-9] The crystals are grown by the Edge-defined Film-fed Growth (EFG™) technique and are known by their CLASS® brand (Clear Large Aperture Sapphire Sheets), see Figure 1. Crystal growth developments this past year has lead to new products with CLASS225® now available in thicker sizes (up to 22 mm), and CLASS400® sheets available up to 400 mm wide. The largest application of these windows, in addition to transparent armor, has been in the aerospace industry for VIS-MWIR optical windows, where the intrinsic strength, hardness (resistance to erosion), chemical resistance, and high optical transmission make sapphire a highly desirable material. Of course, these same properties make sapphire a highly desirable material for transparent armor.

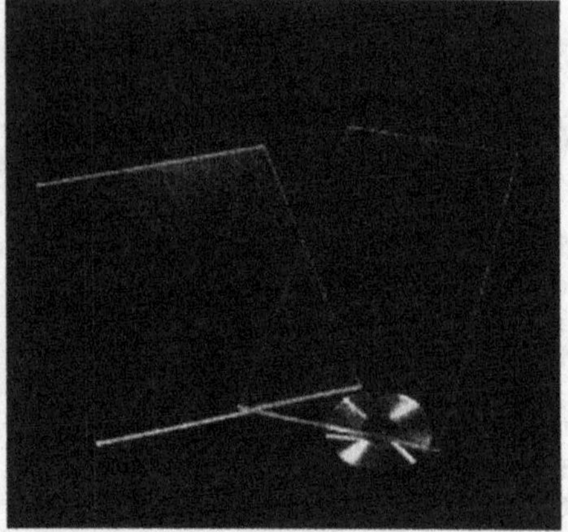

Figure 1. CLASS$^{300®}$ and CLASS$^{225®}$ sheets as grown (left) and after polishing (right).

Traditional transparent armor systems use laminated glass with a polymer backing to stop ballistic projectiles. These systems can be quite thick and heavy due to the amount of glass needed to stop high powered projectiles. As threats become tougher to defeat and specific threats become harder to stop with traditional laminated glass, ceramic composites are the leading solution to reduce thickness and weight, while improving protection and transparency. This is especially true for armor piercing rounds.

The four leading strike-face materials used in armor are glass, sapphire (Al_2O_3), aluminum oxynitride [$AlN_x(Al_2O_3)_{1-x}$], and spinel ($MgAl_2O_4$). Figure 2 contains pertinent material properties for each of these materials.[10-12] Sapphire has the best fracture toughness, flexure strength, thermal conductivity, hardness and modulus of elasticity (Young's modulus) of all the strike face materials. The density of sapphire is almost double glass, but only slightly higher than the other ceramic-based strike-face materials. Based on the mechanical properties, sapphire is expected to have the best ballistic performance of all the strike-face materials for transparent armor.

BALLISTIC TESTING

The sapphire transparent armor composites are composed of a sapphire sheet on the strike face and the sapphire is bonded to glass using conventional interlayers, such as polyvinyl butyral, or other polymers. Figure 3 shows a photograph looking through a laminated window (there is a seam in the middle of this particular window). While the exact interlayer and composition of each window is proprietary, all samples have the sapphire strike face, interlayers of glass, and a final polymer layer on the backside of the window. Ballistic tests followed NATO STANAG 4569 unless noted.[13] Due to the existence of multiple test standards, we will not report on meeting specific standards (although, Saint-Gobain can provide this for commercial orders), but rather the projectiles and velocities within each test. Ballistic testing was completed either at Saint-Gobain Sully, or commercial ballistic laboratories. Projectile velocity measurements were obtained with infrared screens and counter chronographs.

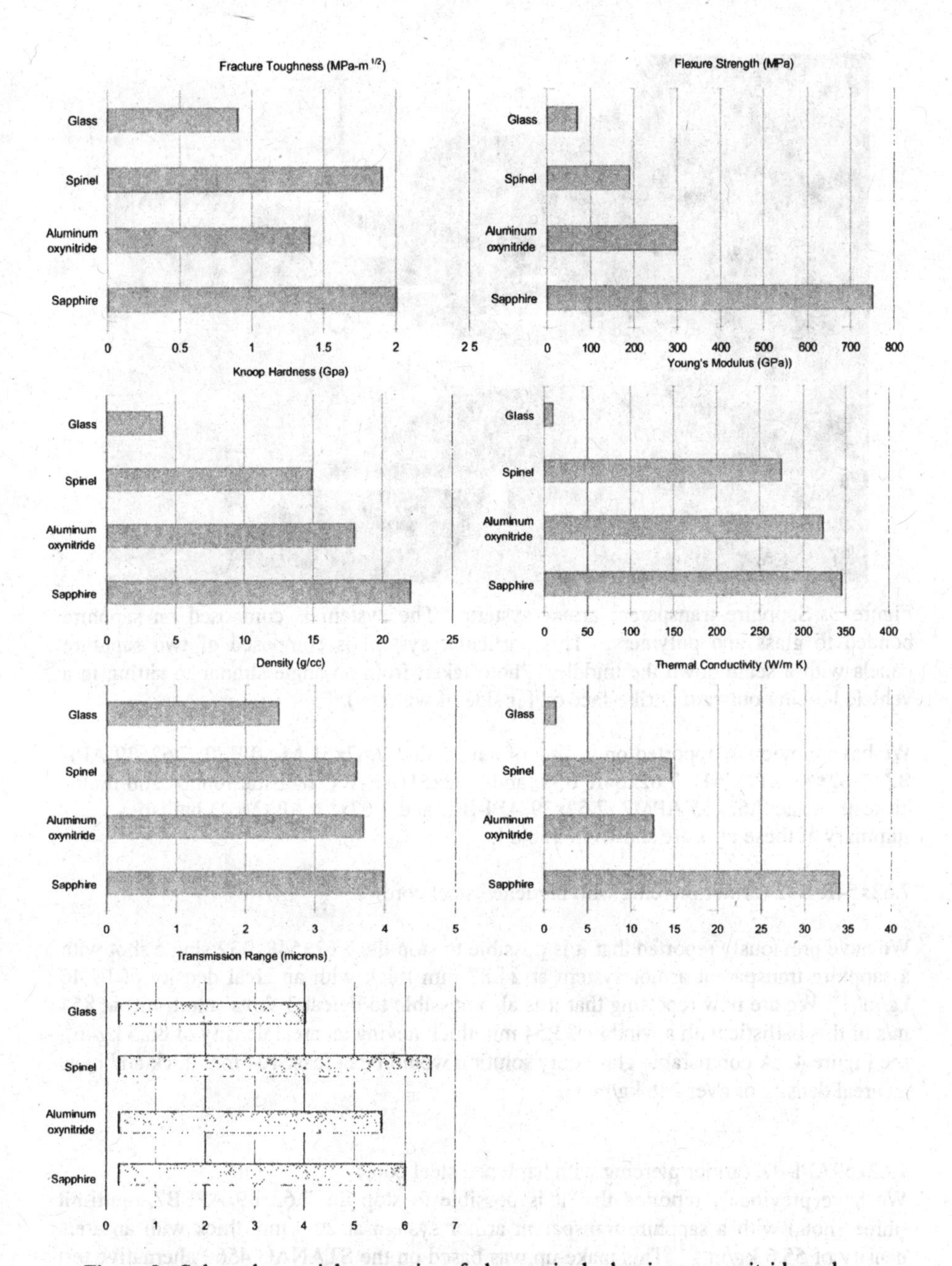

Figure 2. Selected material properties of glass, spinel, aluminum oxynitride and sapphire[10-12]

Figure 3: Sapphire transparent armor system. The system is composed on sapphire bonded to glass and polymers. This particular system is composed of two sapphire panels with a seam down the middle. Photo taken from an angle similar to sitting in a vehicle looking outward (strike-face on far side of window)

We have previously reported on testing of single shot 7.62x51 M-80 Ball, 7.62x39 API-BZ, 7.62x51 AP(M61), 7.62x54R B32, and 7.62x51 AP-WC ballistic rounds and multi-hit tests of the 7.62x63 APM2, 7.62x39 API-BZ, and 7.62x51 AP (M61) ballistics.[14.] A summary of these tests are shown in Table I.

7.62x54R B32 (armor piercing with hardened steel core)

We have previously reported that it is possible to stop the 7.62x54R B32 single shot with a sapphire transparent armor system at 24.87 mm thick with an areal density of 67.46 kg/m^2.[14] We are now reporting that it is also possible to defeat 3 shots (multi-hit) at 854 m/s of this ballistic with a window 33.54 mm thick having an areal density of 86.5 kg/m^2, see Figure 4. A comparable glass-only solution would be at least 104 mm thick and have an areal density of over 248 kg/m^2.

7.62x39API-BZ (armor piercing with hardened steel core)

We have previously reported that it is possible to stop the 7.62x39 API-BZ multi-hit (three shots) with a sapphire transparent armor system at 20.8 mm thick with an areal density of 55.6 kg/m^2.[14] This make-up was based on the STANAG 4569 alternative test for transparent armor.[13] This test is used for transparent armor, as tougher requirements for transparent materials may be impractical due to weight, geometric or human factor constraints. The use of a sapphire front strike face can improve on these constraints due to its high hardness and strength. Figure 5 shows an example of the multi-hit pattern

Table I: Summary of previously reported ballistic tests using Sapphire Transparent Armor.[14] ID# 1-8 are single shot with ID 9-11 being multi-hit.*

ID	Sample Size (mm)	Thickness (mm)	Areal Density kg/m2	Projectile	Projectile Velocity m/s	Penetration
1	150	21.1	52.12	7.62x51 M-80 Ball	835	Partial
2	150	21.1	52.12	7.62x39 API-BZ	776	Partial
3	150	21.1	52.12	7.62x51 AP(M61)	768	Partial
4	150	21.1	52.12	7.62x51 AP(M61)	846	Complete
5	150	29.4	72.78	7.62x54R B32	857	Partial
6	150	24.87	67.46	7.62x54R B32	864	Partial
7	150	43.78	159.21	7.62x51 AP-WC	917	Partial
8	150	46.3	169.11	7.62x51 AP-WC	921	Partial
9	305	41.85	107.97	7.62 x 63mm APM2	Shot 1 Shot 2 Shot 3	Partial Partial Partial
10	305	20.78	55.6	7.62x39 API-BZ	Shot 1:715.9 Shot 2: 712.7 Shot 3: 718.8	Partial Partial Partial
11	305	24.78	65.6	7.62 x 51 AP(M61)	Shot 1:842.6 Shot 2: 837.8 Shot 3: 837.2	Partial Partial Complete

* Partial penetration is a passing result. Complete penetration is a failing result.

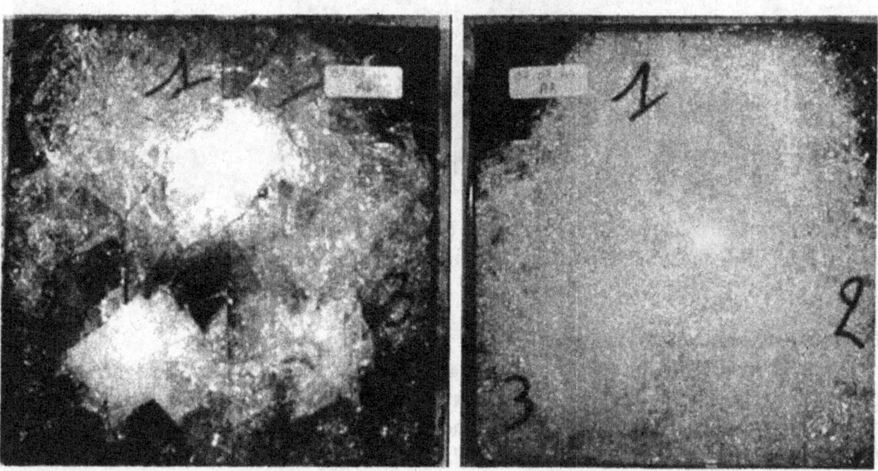

Figure 4: Sapphire transparent armor system (305 x 305 mm) defeating multi-hits with 7.62x54R B32. Left: Strike Face, Right: rear face. Numbers correspond to the order of impact.

(four shots) for the STANAG 4569 test. Figure 6 shows a sapphire transparent armor system (228x330mm) at a thickness of 36 mm with an areal density of 89 kg/m2 that can defeat this tougher STANAG test.

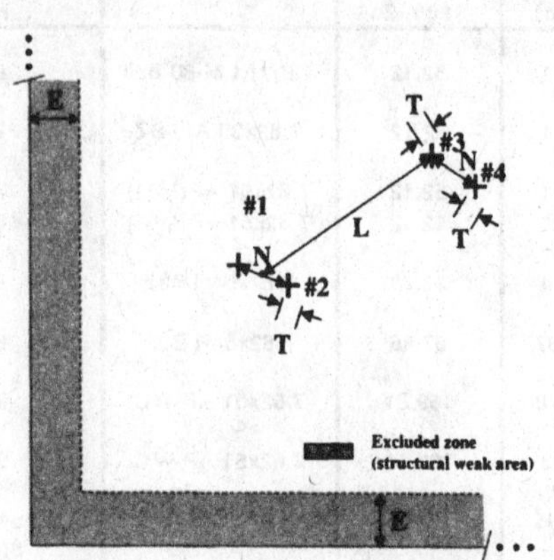

Figure 5 – Example of multi-hit shot pattern from STANAG 4569.[13] The distance between shot #1 and #2 and between shots #3 and #4, N, is 25 mm. The distance between the groups, L, is 100 mm.

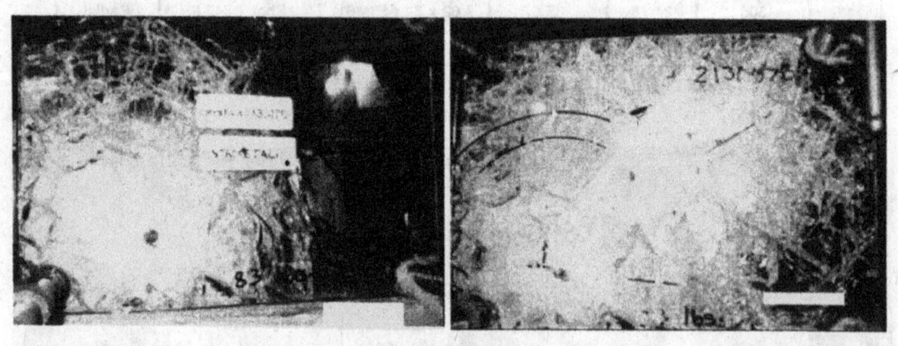

Figure 6: Sapphire transparent armor system (228x330mm) defeating 4 shot multi-hit with 7.62x39 API BZ. Left: Sample after first two shots. Right: sample after all four shots.

Fragmented Simulated Projectiles (FSP)

FSP rounds are used to simulate an artillery threat on armor. While there are no perfect projectiles to simulate improvised explosive devices (IED), FSP rounds are one of the methods used to determine how an armor may stand up to such a threat. There are two sizes of FSP rounds called out in STANAG 4569: 12.7mm and 20mm diameter. FSP are to be made from cold rolled steel and have a hardness value (HRC) of 30 ± 2. The shape of the projectile may be found in Figure 7.

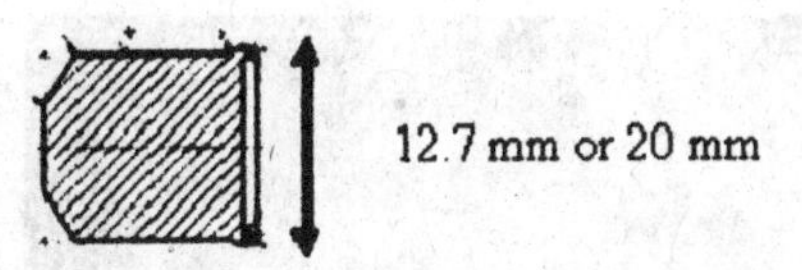

Figure 7: Shape of a Fragmented Simulated Projectile.[13]

Using a 228x330mm sapphire transparent armor system, we defeated three shots (multi-hit) of the 12.7 mm FSP shot at 420 m/s, shown in Figure 8. This composite had a thickness of 47.7 mm and an areal density of 114.8 kg/m^2. Surprisingly, this sample showed little to no backside deformation after shooting.

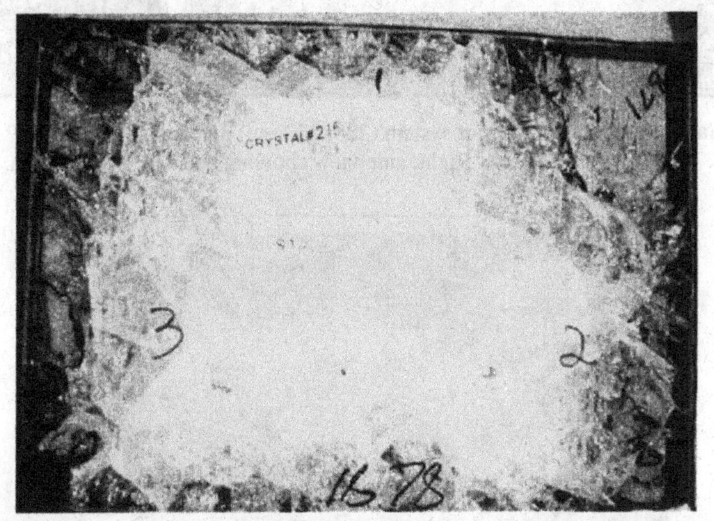

Figure 8: Sapphire transparent armor system defeating three shots (multi-hit) of the 12.7 mm FSP shot at 420 m/s.

The next FSP test was a 20 mm FSP single shot at 780 m/s. It was possible to defeat this threat level using a composite with a thickness of 51.7 mm at an areal density of 124.8 kg/m2, see Figure 9. Note the much higher damage area and large backside deformation due to the increased size and speed of this FSP.

The final FSP threat level tested was a 20 mm FSP shot at 960 m/s. It was possible to defeat this threat level using a composite with a thickness of 93 mm at an areal density of 245 kg/m2, see Figure 10. There was little to no backside deformation on this sample.

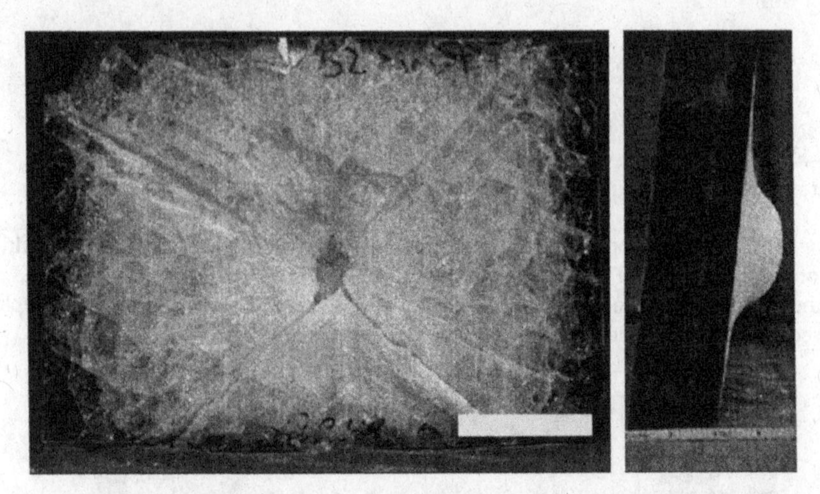

Figure 9: Sapphire transparent armor system (228x330mm) defeating a 20 mm FSP shot at 780 m/s. Left: front strike face. Right: side view showing backside deformation.

Figure 10: Sapphire transparent armor system (305x305mm) defeating a 20 mm FSP shot at 960 m/s.

14.5x 114 API/B32 (hardened steel armor piercing core)
It is reported that it is possible to defeat the 14.5 x 114 API/B32 single shot at a velocity of 911 m/s. The sapphire transparent armor system was 93 mm thick with an areal density of 245 kg/m^2. The sapphire transparent armor system after testing is shown in Figure 11.

Figure 11: Sapphire transparent armor system after being shot with a 14.5x114 API/B32 at 911 m/s.

Seamed Sapphire Panels

While it is possible to make sapphire panels today up to 400 mm wide and up to 600 mm long, there are some instances where sapphire panels larger than this are necessary. One method of making larger transparent armor windows is to bond two or more sapphire panels, leading to a 'seam' in the final composite. While the seamed area is quite small compared to the surface area of the window, it was prudent to test how the seam withstood ballistic threats. The ballistic resistance of the seams were tested under multi-hit conditions.

Two 228x228mm sapphire panels were bonded next to each other to make a 228x456mm sapphire front strike face. A photograph showing an example of this composite is in Figure 3. When looking through the panel, the visual line due to the seam is surprisingly small, a consequence of the high index of refraction in the seamed layer and from the seam being present only in the front strike face, and throughout the thickness of the armor. The completed sapphire transparent armor system had a final thickness of 36 mm and an areal density of 89 kg/m^2.

The first test was a single shot on the seam of a 7.62x39 API/BZ and a single shot of a 7.62x54R B32, see Figure 12. Both samples behaved very similar to a monolithic piece of sapphire, with no visual differences observed and the seam did not appear to widen or move. Additional multi-hit testing (Figure 13) was performed by putting shots two and three 120 mm from the first to form an equilateral triangle showed that both samples still performed similar to a monolithic sample and passed the test with these thicknesses and areal densities. To further test the ballistic resistance of the seam, another sample (Figure 14) was shot twice with 7.62x39 API/BZ rounds—this time both shots were on the seam. Once again the panel passed the test.

Comparison of Sapphire Transparent Armor to Traditional Glass Armor

As we have demonstrated above, sapphire transparent armor can defeat a variety of threats, including multi-hit rounds. The advantage of using sapphire as a strike face is that it has high flexure strength, high hardness, and good toughness. These properties

Figure 12: A seamed sapphire transparent armor system after shooting with the 7.62X39 API/BZ (left) and 7.62x54R B32 (right).

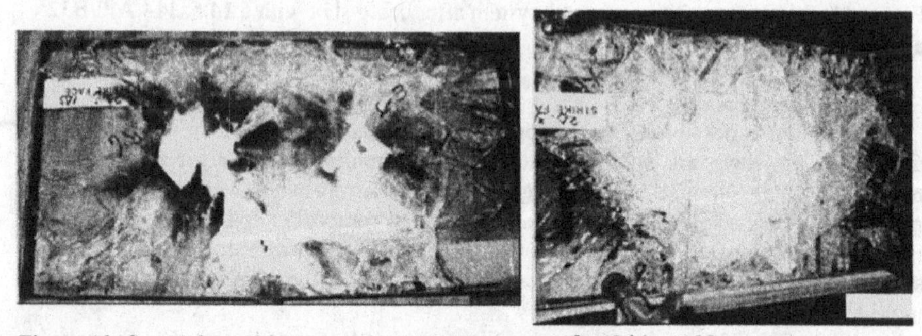

Figure 13: Seamed sapphire transparent armor system after 3 hits (120 mm space triangle) of the 7.62X39 API/BZ (left) and 7.62x54R B32 (right). In both samples, the first shot was on the seam.

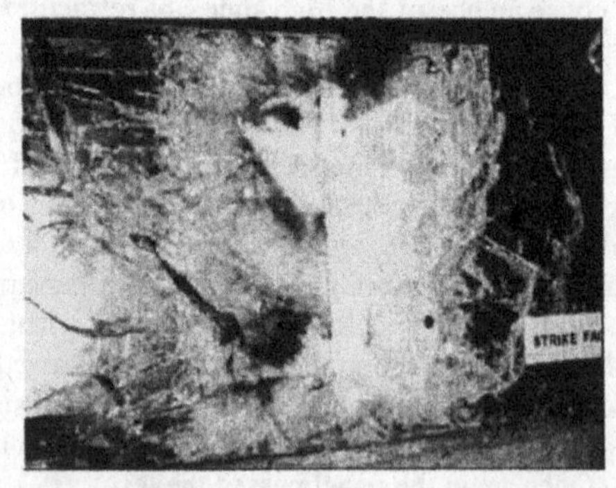

Figure 14: Seamed sapphire transparent armor system with a second shot on the seam of a 7.62x39 API/BZ 120 mm under the first.

contribute to the destruction of the hard core in armor piercing projectiles, thus greatly reducing the armor piercing capabilities of the threat. Table II shows the comparison between sapphire transparent armor and traditional glass armor to defeat two different armor piercing rounds. It is clear that the sapphire transparent armor represents a step-change in technology for transparent armors, with thickness savings up to 68% and weight savings up to 65% when compared to traditional glass laminate armor.

Table II: Comparison of Sapphire Transparent armor to Glass armor

Threat	# shots	Glass thickness (mm)	Sapphire armor thickness (mm)	Thickness savings with sapphire armor	Glass areal density (kg/m^2)	Sapphire armor areal density (kg/m^2)	Weight savings with sapphire armor
7.62x39 API/BZ	3	58	20.8	64%	133	56	58%
7.62x54R B32 API	3	104	33.5	68%	248	86	65%
7.62x54R B32 API	1	55	24.8	55%	115	67.5	41%
14.5 x 114 API/B32	1	150	93	38%	360	245	32%

CONCLUSION:

Saint-Gobain has developed a commercial ceramic transparent armor system based on EFG™ sapphire. The sapphire transparent armor system can reduce the thickness of a window by up to 68% and reduce the weight by up to 65% over traditional glass transparent armor. We have presented sapphire transparent armor solutions that can defeat a variety of multi-hit threats, including 7.62x39 API-BZ, 7.62x54R AP B32, 7.62x63 APM2, and 12.7 mm fragmented simulated projectiles (FSP). Additionally we have presented solutions that can defeat single shot 20 mm FSP and 14.5x114 API B32. We have also investigated the ability to make a mosaic-type window, combining two sapphire windows side by side on the strike face. We found that this system performed very similar to an armor system that has a monolithic sapphire panel on the strike face. This mosaic approach may be used to make larger transparent armored windows.

REFERENCES

1. C. D. Jones, J. B. Rioux, J. W. Locher, H. E. Bates, S. Zanella, V. Pluen, M. Mandelartz, *Large-Area Sapphire for Transparent Armor*, Proceedings of the 30th International Conference on Advanced Ceramics and Composites, The American Ceramic Society, January 2006.

2. C. D. Jones, J. B. Rioux, J. W. Locher, H. E. Bates, S. Zanella, V. Pluen, M. Mandelartz, *Large-Area Sapphire for Transparent Armor*, Amer. Cer. Soc. Bull., Vol. 85, No. 3 March 2006.

3. H. E. LaBelle, *EFG the invention and application to sapphire growth*, Journal of Crystal Growth, Vol. 50, 8 (1980).

4. J.W. Locher, H.E. Bates, S.A. Zanella, E.C. Lundstedt, and C.T. Warner, *The production of 225 x 325 mm sapphire windows for IR (1 to5 μm) application*, Proc. SPIE Vol. 5078, Window and Dome Technologies VIII. Randal W. Tustison, Editor, 2003, pp.40-46

5. J. W. Locher, J. B. Rioux, and H. E. Bates, *Refractive Index Homogeneity of EFG A-plane Sapphire for Aerospace Windows Applications*, 10th DoD Electromagnetic Windows Symposium, Norfolk VA, 18-20 May 2004.

6. H. E. Bates, C. D. Jones, and J.W. Locher, *Optical and crystalline characteristics of large EFG Sapphire Sheet*, Proc. SPIE Vol. 5786, Window and Dome Technologies and Materials IX, Randal W. Tustison, Editor, May 2005, pp. 165-174

7. J. W. Locher, H.E. Bates, C. D. Jones, and S. A. Zanella, *Producing large EFG sapphire sheet for VIS-IR (500-5000 nm) window applications*, Proc. SPIE Vol. 5786, Window and Dome Technologies and Materials IX, Randal W. Tustison, Editor, May 2005, pp. 147-153

8. C. D. Jones, J. W. Locher, H.E. Bates, and S. A. Zanella, *Producing large EFG sapphire sheet for VIS-IR (500-5000 nm) window applications*, Proc. SPIE Volume 5990, 599007 (2005).

9. V. A. Tatartchenko, *Sapphire Crystal Growth and Applications*, in Bulk Crystal Growth of Electronic, Optical, and Optoelectronic Materials, edited by P. Crapper, pp 299-338, John Wiley & Sons (2005)

10. D.C. Harris, *Materials for infrared windows and domes, properties and performance*, SPIE Optical Engineering Press, Bellingham WA, 36 (1999).

11. Surmet Corporation Company Literature #M301030

12. Saint-Gobain Crystals Product Literature

13. NATO STANAG 4569; Procedures for evaluating the protection levels of logistic and light armoured vehicles for KE and artillery threats.

14: C. D. Jones, J. B. Rioux, J. W. Locher, H. E. Bates, S. Zanella, V. Pluen, M. Mandelartz, *Large-Area Sapphire for Transparent Armor*, Proceedings of the 31th International Conference on Advanced Ceramics and Composites, The American Ceramic Society, pp113-124 January 2007.

DEFECT FREE SPINEL CERAMICS OF HIGH STRENGTH AND HIGH TRANSPARENCY

Juan L. Sepulveda, Raouf O. Loutfy, and Sekyung Chang
Materials and Electrochemical Research (MER) Corporation
Tucson, Arizona 85706, USA

ABSTRACT

This paper describes recent advances on the development of low cost, high strength, transparent magnesium aluminum spinel windows and domes for armored vehicles, aircrafts, and missiles. The novel spinel material exhibits high optical and IR transparency in the 0.3 – 5.5 μm wavelength, is very resistant to abrasion, with density higher than 99.95% of theoretical, with very fine and uniform grain size, and flexural strength of 300 MPa. Its mechanical properties are several times greater than that of glass and make it a leading candidate for use as a transparent armor and window material. The novel spinel described in this paper is superior to conventional available spinel materials. Improved performance derives from the use of an advanced transient sintering aid technology that leaves no traces in the final product, and flexible green forming and densification achieved by high temperature sintering of high purity spinel powder. Unlike previous efforts to produce transparent spinel by conventional mixing or granulation during powder preparation, resulting in inhomogeneous spinel with exaggerated grain growth and hazy and/or opaque regions, the novel powder preparation technology ensures a previously unattainable uniformity in sintering aid dispersion that allows the densification of highly transparent and uniform spinel windows. The production of discs, lenses, curved windshields, domes, and rectangular windows larger than 12"x16" in large numbers, exhibiting no optical defects, high degree of transparency, using conventional ceramic powder compaction technology or net shape green body formation is described. Spinel transmission and mechanical properties are presented.

INTRODUCTION

Transparent spinel ceramics intended for windows and domes are of great interest for applications where high transparency in the UV-VIS and IR wavelength range is desired but at the same time shielding protection and abrasion resistance are also requirements. MER has developed and scaled up transparent magnesium aluminate spinel technology for the production of transparent windows and windshields intended for aircraft sensing equipment, HMMWV, MRAP, FMTV, HEMTT , other military vehicles, and other armor applications, and at an affordable cost[1,2].

Spinel technology has been under development by several groups of investigators[3-22]. MER emphasis during the last three years was to introduce refinements based on nano-technology and to further scale up, improve, and lower the cost of spinel parts by pursuing the development of plant scale technologies applicable in an industrial environment. After completion of two Phase I SBIR programs[1,2] and other self-funded development efforts, MER has accomplished such an objective. The manufacturing processes consisting of green forming, hot pressing (HP), hot isostatic pressing (HIP), and polishing were developed, scaled-up, and simplified while keeping the quality and performance of the product and at the same time while remarkably reducing production cost. Polished spinel parts have been produced exhibiting high degree of transparency. This was accomplished by the remarkable elimination of intergrain and intragrain porosity, by reducing all contamination, and by keeping the grain size below 30 μm for increased strength. This outcome is a unique characteristic of the MER developed process that is not commonly found in the existing market spinel offering. Two US Patents are pending. The timeline for MER spinel technology development and scale up is graphically depicted in Figure 1. After technology development and optimization during the first two years, current production efforts have concentrated producing several parts per sintering run in sizes up to 12.4"x16.4"x0.6". The production of up to four such parts per hot press run has been successfully demonstrated by MER in 2008.

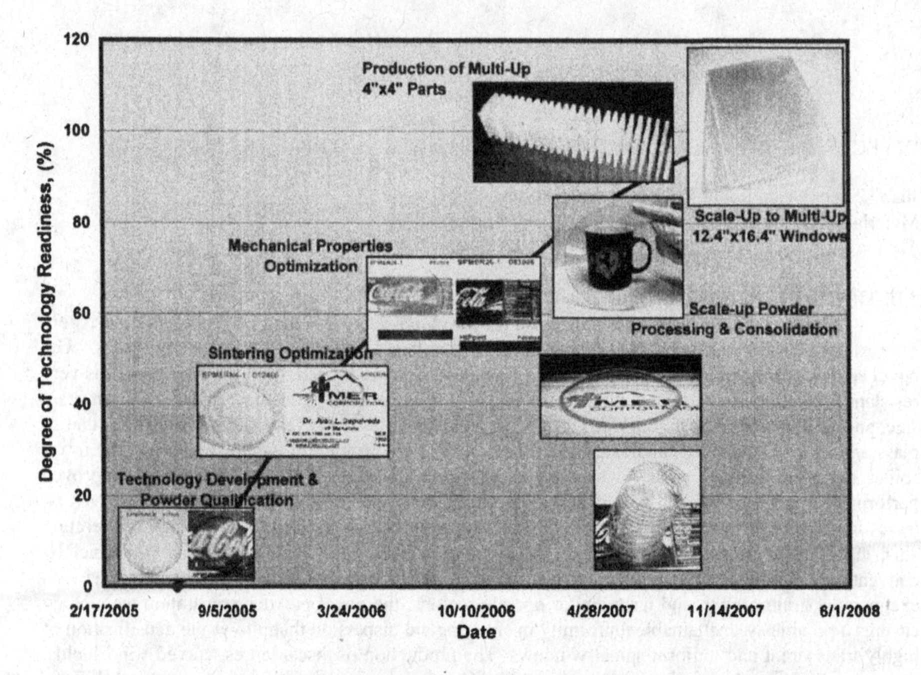

Figure 1. Timeline for MER Spinel Technology Development and Scale-Up

SPINEL TECHNOLOGY

As shown in Figure 2 and Figure 3, transparent, defect-free, spinel windows as large as 16"x12" have been produced using the recently developed technology which is well fitted for large scale industrial production. The production of multiple spinel parts per sintering run of the highest transparency was demonstrated.

Typical MER spinel material properties are summarized in Table I. Some of these properties could still be optimized. For example, strength is one property that could still be improved.

Figure 2. 12.4"x16.4" Sintered Spinel Windows, top to bottom: 0.45", 0.45", and 0.60" thick.

Table I. MER Spinel Material Properties

Property	Value
Chemical Composition	$MgAl_2O_4$
Crystal Structure	Cubic, spinel, polycrystalline
Density	3.58 g/cc
Melting Point	2150°C
Coeff. Therm. Exp.	6.34 ppm/°K
Therm. Cond.	10 W/mK
Elastic Modulus	277 GPa
Poisson's Ratio	0.28
Hardness	1600 Kg/mm^2
Flexural Strength	300 MPa
Fracture Toughness	1.7 MPa-m$^{1/2}$
Transmission	0.3 - 5.5 μm
Forward Scatter	<0.2%

Recently, MER has also developed and demonstrated the feasibility of producing transparent magnesium aluminum spinel domes for missiles at an affordable cost using low pressure near-net-shape green forming technology[2]. Transparent, defect-free, mechanically generated and polished, 4" in diameter, 3.5" radius, spinel domelets were produced as shown in Figure 4 and 5. The developed technology is fully capable for the production of hemispherical domes, curved windshields, and other complex geometry parts. The production of high quality, defect-free, domes is accomplished using low-pressure sintering of the green formed dome blanks. Additional spinel technology advances achieved recently are the production of large size multiple spinel parts of the highest transparency per sintering run, and the reduction in vacuum requirements during sintering. One of the main advantages of the new technology developed by MER consists of an improved and more efficient spinel powder preparation that results in higher capacity, lower powder contamination, and higher powder utilization as compared to standard powder preparation technology. A major advantage of the new process is the very high efficiency of spinel powder utilization achieved. Better than 99.5% powder yield was attained during powder preparation using the novel powder preparation process. After low pressure sintering, 100% part yield was obtained after performing several consecutive runs. All these process advantages will result in lower production cost.

Figure 3. 9"x4"x0.27" Polished Spinel Window.

Figure 4. 4" Diameter, 3.5" Inner Radius, Spinel Dome Concave Face Up.

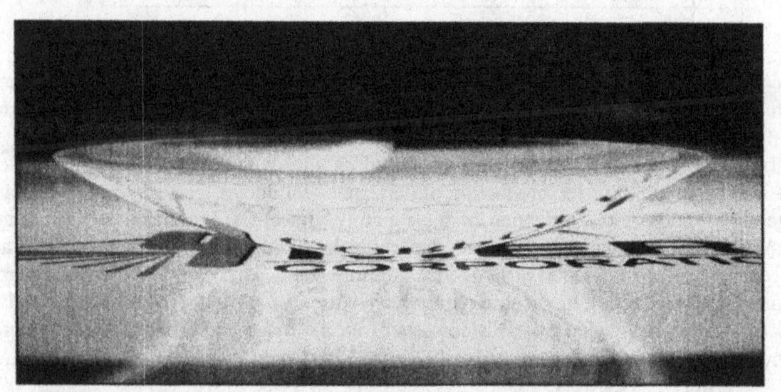

Figure 5. 4" Diameter, 3.5" Inner Radius, Spinel Dome Convex Face Up.

A very important feature of the MER process is the use of a transient additive package consisting of binders and sintering aids that effectively participate at different stages of the densification process but then they are eliminated leaving no contamination in the sintered spinel body.

The powder preparation process makes use of intensive particle de-agglomeration under well controlled ultra-pure conditions using nano-technology to attain high degree of ingredients dispersion. Using nano-sized powders and nano-technology procedures the sinterability is highly enhanced while purity is preserved to obtain highly transparent, fine grained, inclusion free sintered spinel parts.

The prepared powder can be agglomerated using conventional binders or a specially designed binder system that leaves no residue after sintering. Powder agglomeration can be accomplished using conventional spray drying or freeze drying. Green forming can be accomplished by conventional cold isostatic pressing (CIP) or cold dry pressing to produce near-net-shape parts.

Corresponding sintering profile has been refined and optimized to attain low pressure sintering of the domes. Special attention was given to temperature treatment for phase formation and elimination of all contaminants in the spinel powder. The maximum sintering temperature and soak time were optimized to achieve 99.5+% of theoretical density but at reduced grain size which resulted in flexural strength that reached 300 MPa. Parts of the highest transparency with no stains, blemishes, or optical defects were produced. Transmission in the 0.3-5.5 μm wavelength range reached close to 90%.

Figure 6. 4" Diameter Green Spinel Domes of Different Thickness.

Improvements achieved for sintering were complemented with process improvements which allowed for the production of several parts per run. The technique was demonstrated during the sintering of up to 4 window or disc shaped parts per run.

By using close-to net-shape green forming prior to sintering fully dense sintered domes (>99.5% of Th.D.) were produced. Typical dome green preforms are shown in Figure 6. After powder preparation, HP, HIP, and polishing, highly transparent dome were produced having uniform optical and mechanical properties in absence of grains of exaggerated size. Densification is accomplished by using low pressure forging at temperature of 1500°-1700°C. Unlike previous efforts to produce transparent spinel by mixing and grinding the sintering aid that produces inhomogeneous spinel with exaggerated grain growth, and hazy and/or opaque regions, MER's unique powder preparation technology ensures a previously unattainable uniformity in sintering aid dispersion that allows the consistent densification of highly transparent and uniform spinel domes as shown in Figure 4 and Figure 5.

The high fired density obtained under these conditions is believed to be caused by the formation of fine lamellar microstructure in the solidified vehicle, inhibiting porosity formation. Experimental efforts demonstrated the feasibility of producing transparent, defect-free, spinel domes 4" in diameter, 3.5" radius, 0.18" thick, using an innovative green forming process specially fitted for large scale industrial production. Several different powder lots were qualified and successfully used. Gathered experimental results clearly demonstrated that green net shape forming is mandatory for the production of complex net shape sintered domes. This will insure overall dome morphology and dome wall thickness while at the same time insuring density uniformity through out the dome body. An intelligent graphite die design, that makes use of these attributes during sintering, makes the production of domes and hemispheres feasible and very successful.

During the course of this project, several 4" in diameter domes were produced resulting in 100% yield. Another major accomplishment was the refinement developed for the sintering profile during high temperature, low pressure HP (forging) of the domes. After several experimental runs it was concluded that low pressure HP sintering practice is mandatory for the large scale, low cost, production of spinel domes. The technology can be easily scaled up for larger size parts as large missile domes, protective windows for aircrafts, armored vehicles, armor, protective goggles, ruggedized computers and displays, and laser weapons needed by the defense industry while commercial applications such as fixed and portable point of sale (POS) terminals, watch crystals, vapor lamp tubes, firefighter and police face shields, other applications where a rugged, tough, scratch, resistant, transparent material is required will also benefit from this technology.

TRANSMISSION

Spinel transmission capabilities were measured in the UV-VIS range of 200 to 900 nm and in the Mid IR range 2272 to 8000 nm. For the wavelength range 200 nm to 900 nm a Perkin Elmer Lambda 3B UV/VIS Spectrophotometer was utilized. A baseline scan without samples was performed and the corresponding deviation from 100% was calculated and used to adjust the sample measured readings. Fourier-Transform Infra-Red (FTIR) spectroscopy for the wavelength range 2272 nm to 8000 nm was conducted in a Perkin Elmer 1600 Series FTIR Spectrophotometer. Samples were placed directly in the laser beam after a background spectrum was taken with no sample in the beam. A total of 16 scans were averaged to yield the final spectrum. Perkin-Elmer Spectrum processing software was used to subtract the background, and smooth and format the spectral data.

Typical transmission spectra are shown in Figure 7 and Figure 8. High degree of transparency was obtained as shown in these examples. Parts that were fully polished and HIPped resulted in higher degree of transparency. Chemical mechanical polishing (CMP) also improved transparency as shown in Figure 8.

A very good match in between UV-VIS and Mid-IR measurements was obtained as shown in Figures 7, and Figure 8. The transmission in between 900 nm and 2272 nm was extrapolated. Transmission data obtained so far seem to indicate CMP polishing provided better transmission than mechanical polishing for the 5-7 μm wavelength.

STRENGTH

Spinel strength, as with most other ceramics, increases as the grain size decreases. Spinel strength improvement was accomplished by reducing grain growth by adjusting the soak time and/or the sintering temperature. Within limits, a reduction in grain size resulted in improved strength but at the same time the density was kept high. The relationship in between grain size and retention time at sintering temperature (soak) was pursued during hot pressing (HP). 1⅛" spinel parts were sintered for ½ hr, 1 hr, 2 hr, and 4.5 hr. Figure 9 shows the variation in grain size versus the sintering time.

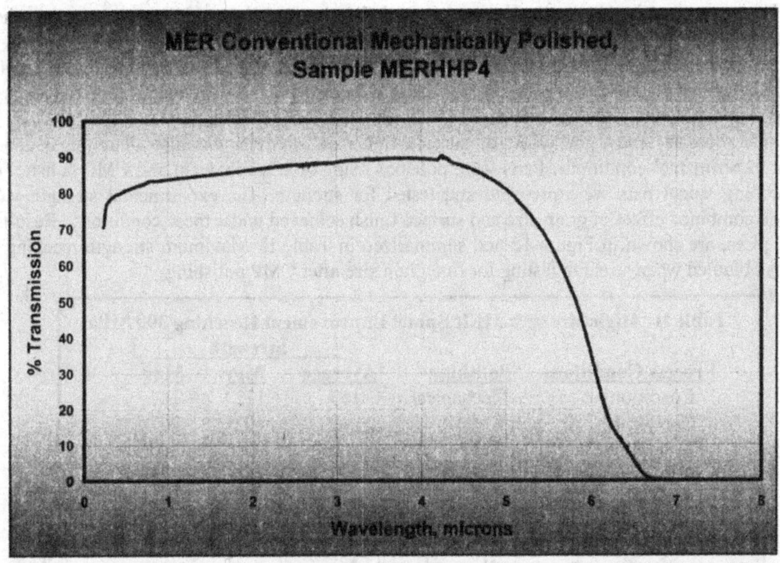

Figure 7. UV-VIS and IR Transmission Characteristics for a Mechanically Polished Spinel Disc after HP and HIP.

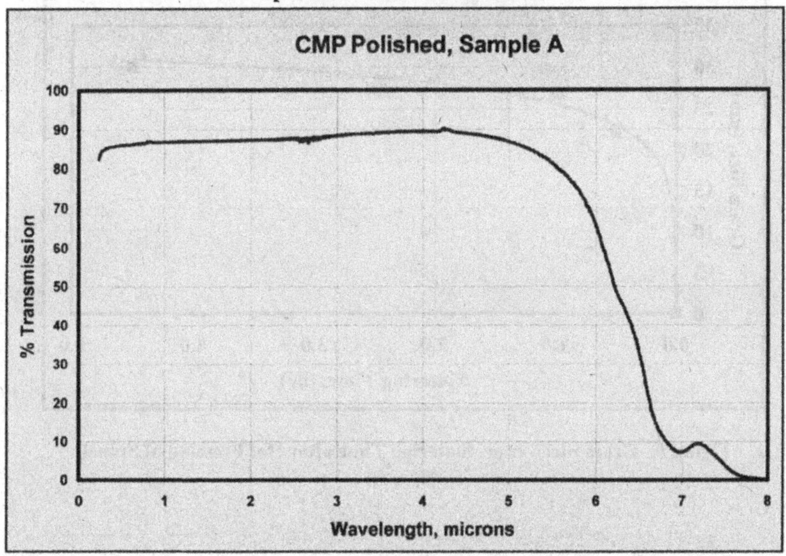

Figure 8. UV-VIS and IR Transmission Characteristics for a CMP Polished Spinel Disc after HP and HIP.

The corresponding Vickers hardness measured at 500 g for 10 seconds was measured as shown in Figure 10. Figure 11 shows the dependency of hardness on grain size. It can be concluded, the higher the hardness, the higher the expected strength. Reducing the soak time at sintering temperature during HP reduced the grain size as expected. Grain size of 28.5 μm obtained after two hours of sintering resulted in maximum hardness (1704 Kgf/mm^2) and resulted in higher strength.

In a separate study, the effect of surface finish on strength was also determined under 4 different experimental conditions. Parts were polished using both mechanical and CMP polish. Four-point bending spinel bars were prepared and tested for strength. The experimental strength values reflect the combined effect of grain size and surface finish achieved under these conditions. Results, as Weibull plots, are shown in Figure 12 and summarized in Table II. Maximum strength reaching 300 MPa was obtained when strength testing for fine grain size after CMP polishing.

Table II. High Strength MER Spinel Improvement Reaching 300 MPa.

		Strength		
Process Conditions	**Polishing**	**Average**	**Min**	**Max**
Conditions 1	Mechanical	153	112	182
Conditions 2	CMP	245	192	267
Conditions 3	CMP	286	256	306
Conditions 4	CMP	213	173	229

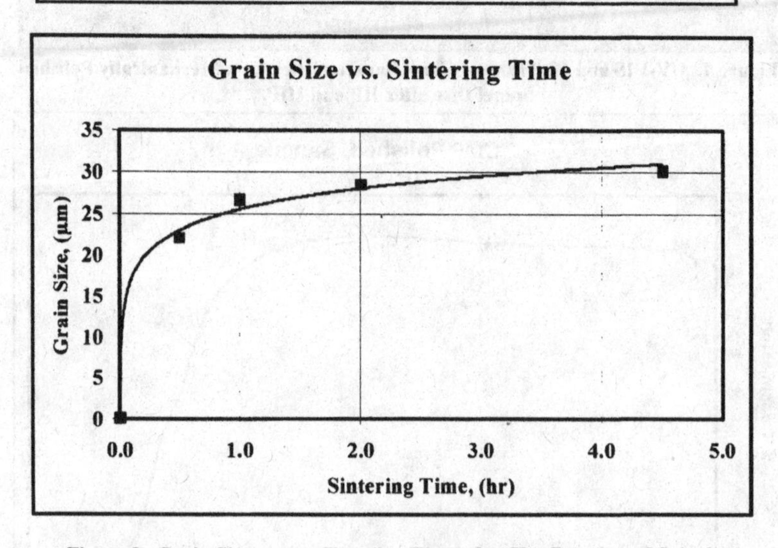

Figure 9. Grain Size versus Sintering Time after Hot Pressing of Spinel.

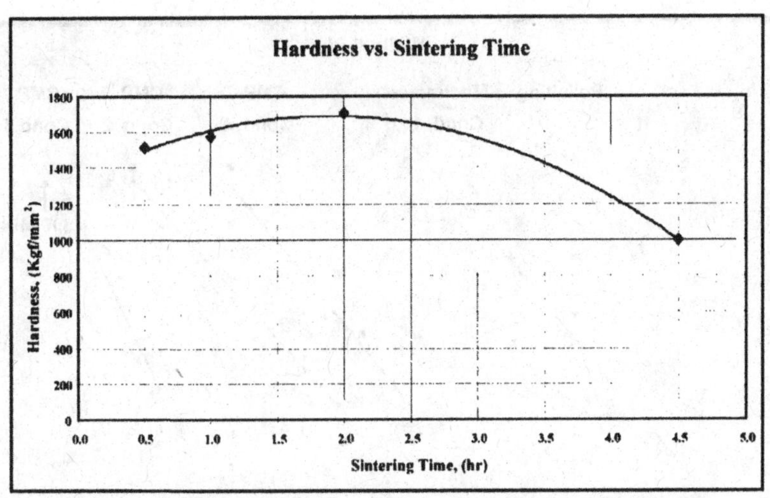

Figure 10. Hardness versus Sintering Time after Hot Pressing of Spinel.

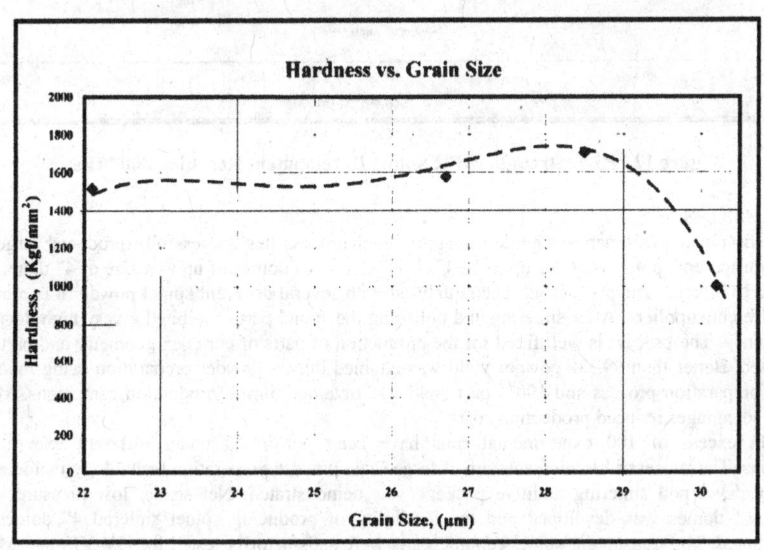

Figure 11. Hardness versus Grain Size after Hot Pressing of Spinel.

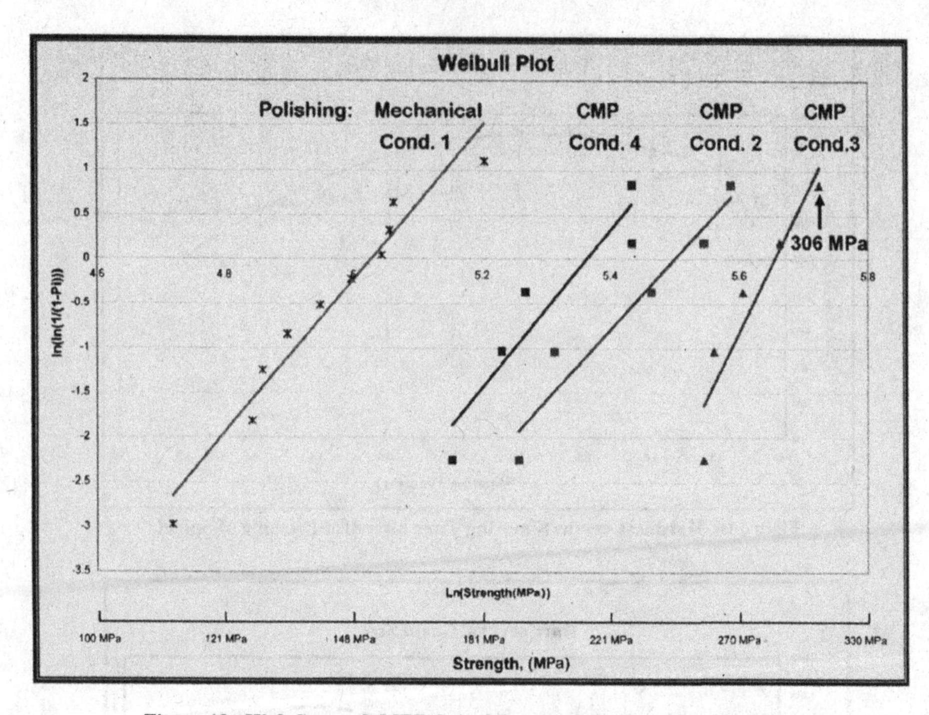

Figure 12. High Strength MER Spinel Development Reaching 300 MPa.

CONCLUSIONS

The novel MER nano-technology based spinel process has successfully produced large size highly transparent spinel windows up to 12.4"x16.4"x0.6" and domes of up to a size of 4" diam., 3.5" radius, 0.18" thick. The process has been qualified with several different spinel powder lots procured from different suppliers. After sintering and polishing the spinel parts exhibited a very high degree of transparency. The process is well fitted for the production of parts of complex geometry and parts that are curved. Better than 99.5% powder yield was attained during powder preparation using the novel powder preparation process and 100% part yield was obtained during production campaigns. MER's process advantages reduced production cost.

In excess of 100 experimental runs have been performed under different experimental conditions. The successful implementation of large scale powder preparation technology including the specially developed sintering additive package was demonstrated. Net shape, low pressure, green forming of domes was developed and the feasibility of producing spinel sintered 4" domes was demonstrated. The transparent spinel ceramic parts were satisfactorily tested for UV-VIS and Mid-IR transmission in the 0.3-5.5 μm wavelength range. High transparency and transmission were obtained by eliminating residual intergrain and intragrain porosity and all sources of potential organic, ceramic, and metallic contamination. The net result was the elimination of inclusions, discoloration, blotches, clouds, and blemishes in the spinel fired body.

It was concluded that green net shape forming is mandatory for the production of complex net shape sintered domes. Low pressure hot pressing is mandatory for the multi-up large scale production of large spinel parts.

ACKNOWLEDGMENTS

This program was performed under the auspices of the U.S. Air Force under SBIR Phase I Contract FA8650-06-M-5507, and the U.S. ARMY under SBIR Phase I Contract W31P4Q-07-C-0080. The characterization support received from the Naval Research Laboratory is also acknowledged. Additional funding was provided by NMIC, Tucson, AZ.

REFERENCES

[1] J. L. Sepulveda, U.S. Air Force, SBIR Phase I, Contract FA8650-06-M-5507, Final Report, March, 2007.
[2] J. L. Sepulveda, U.S. ARMY, SBIR Phase I, Contract W31P4Q-07-C-0080, Final Report, August, 2007.
[3] G. R. Villalobos et al., *J. Am. Ceram. Soc.*, 88 [5] 1321–1322 (2005)
[4] G. R. Villalobos et al., Materials Science and Technology, 2005 NRL Review.
[5] S.S. Bayya, et al., Proc. of SPIE Vol.5786, Bellingham, WA, 2005.
[6] Sellers et al., U.S. Pat. No. 3,768,990.
[7] D. C. Harris, Materials for Infrared Window and Dome, SPIE Press, Bellingham, WA, 1999.
[8] D.C. Harris, Infrared Window and Dome Materials, SPIE Opt. Eng. Press, Bellingham (1992).
[9] J.C. Kirsch, et al., Proc. of SPIE Vol.5786, Bellingham, WA, 2005.
[10] T.J. Mroz, et al., Proc. of SPIE Vol.5786, Bellingham, WA, 2005.
[11] I.E. Reimanis, et al., Proc. of SPIE Vol.5786, Bellingham, WA, 2005.
[12] K. E. Sickafus et al., *J. Am. Ceram. Soc.*, 82 [12] 3279–92 (1999)
[13] Z.S. Rak, Polish Ceramics 2000 Conference, Spala May 29-31, 2000, ECN-RX-00-003.
[14] V. T. Gritsyna, et al, *J. Am. Ceram. Soc.*, 82 [12] 3365–73 (1999)
[15] J-L Huang et al., *J. Am. Ceram. Soc.*, 80 [12] 3237–41 (1997)
[16] R. Cook et al., Proceedings of the Defense and Security Symposium 2005.
[17] A. Ghosh et al., *J. Am Ceram Soc*, 74 [7]1624-30 (1991)
[18] C. L. Patterson, at al., SPIE 45th Int. Symp. on Opt. Sci. and Tech., July 2000, San Diego, CA.
[19] W. Roy et al., Eighth DoD Electr. Windows Symp., April 2000, Colorado Springs, CO.
[20] W. Roy et al., Am. Ceram. Soc. Bull., 52 [4] 372–3 (1973).
[21] C. L. Patterson et al., Ninth DoD EM Windows Symposium, May 2002, Huntsville, AL.
[22] M. Shimada et al., Mater. Lett., 28 [8] 413–5 (1996).

Opaque Ceramics

RECENT RESULTS ON THE FUNDAMENTAL PERFORMANCE OF A HOT-PRESSED SILICON CARBIDE IMPACTED BY SUB-SCALE LONG-ROD PENETRATORS

Jerry C. LaSalvia, Brian Leavy, Herbert T. Miller, Joshua R. Houskamp, and Ryan C. McCuiston
U.S. Army Research Laboratory
Aberdeen Proving Ground, AMSRD-ARL-WM-MD, Aberdeen, MD 21005-5069

ABSTRACT

The dwell/penetration transition velocity and penetration velocities were determined for a commercially-available hot-pressed silicon carbide (SiC) in a baseline target configuration. The targets consisted of SiC cylinders slip-fitted into Ti6Al4V cups and thin Ti6Al4V cover plates welded to the cups. The lateral and back surfaces of the SiC cylinders were well confined in this target configuration, but not pre-stressed. The diameter of the ceramic cylinders was 38.1 mm. Three ceramic thicknesses were evaluated (12.7, 25.4, and 38.1 mm) in this study. Laboratory-scale tungsten-based (W-based) long-rod penetrators (L/D 20, D = 3.175 mm) were fired into the targets at various velocities ranging between 1000 – 1700 m/s. The ballistic interaction between the penetrators and ceramic targets were captured in time using multiple 1 MeV flash X-rays. The resulting X-ray radiographs were used to determine the dwell/penetration transition velocity and penetration velocities. The results show no apparent effect of thickness on the dwell/penetration transition velocity; however, an effect on penetration velocity was noted for the 12.7 mm thick cylinders. Data for this SiC which includes the effect of lateral pre-stress on the dwell/penetration transition velocity and penetration velocity has recently been published [1]. Comparison of the two data sets indicates that there is a large effect of lateral pre-stress on the dwell/penetration transition velocity, but not on penetration velocity. The experimental set-up and results are discussed.

INTRODUCTION

During the last 20 years, a number of fundamental ballistic studies have been conducted on ceramics [1-12]. Some of these studies sought to determine the "intrinsic" penetration resistance of ceramics [1-6], while others examined the influence of specific armor design parameters [7-11]. In addition, several of these studies tried to identify important ceramic failure mechanisms which governed transitions in ballistic behavior [5,9,10,12].

Traditional ballistic performance metrics used to evaluate the efficiency of either armor ceramics (e.g. depth-of-penetration, DOP) or ceramic armors (e.g. V50) do not normally provide the necessary "resolution" for clearly identifying the connections between ceramic attributes (other than density and cost) and ballistic performance. This is due in part to the fact that these metrics are typically the end result of a multitude of complex highly nonlinear mechanical (elastic, inelastic, and failure) phenomena integrated over both space and time. Consequently, they often mask the key ceramic failure mechanisms which dominate overall performance.

Figure 1 is a simplified graphic illustration of possible penetration-time (*p-t*) behaviors for a ductile penetrator impacting a ceramic at three different impact velocities v_1, v_2, and v_3 ($v_1 < v_2 < v_3$). The *p-t* behavior corresponding to v_1, where the penetrator does not penetrate the ceramic over the duration of the ballistic event (t_{total}), would be expected if v_1 is below the critical impact velocity for penetration (i.e. $v^{transition}$). This phenomenon has been called "interface defeat" [1,4,5,8] and $v^{transition}$ referred to as the dwell/penetration transition velocity [11-12]. For the intermediate impact velocity v_2, the *p-t* behavior exhibits both non-penetrating and penetrating phases. The non-penetrating phase has been referred to as "dwell" [11-12] since the penetrator remains at the top surface of the ceramic for a length of time t_{dwell} before it penetrates into the ceramic. The *p-t* behavior corresponding to v_3 would be expected if v_3 is much higher than $v^{transition}$, such that dwell does not occur to an appreciable extent.

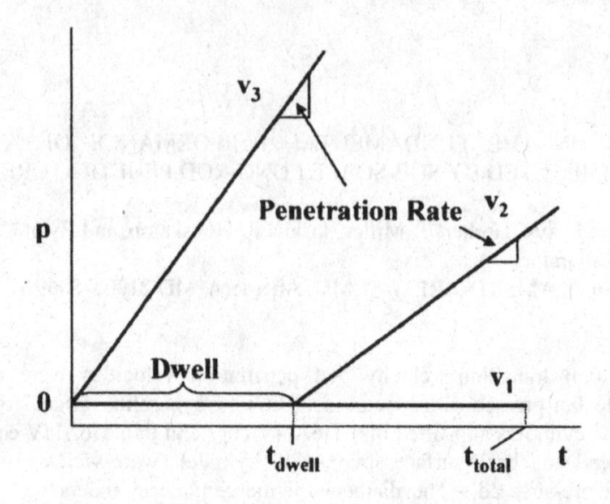

Figure 1. Graphic illustration of penetration histories for three different impact velocities v_1, v_2, and v_3 ($v_1 < v_2 < v_3$).

As might be expected, the occurrence of dwell, as well as the dwell/penetration transition velocity and penetration velocity, are dependent upon material, armor design, and projectile parameters, and consequently, can vary significantly [1,4,5,8,11]. How and why they vary is important knowledge for the development of improved ceramics. In this paper, results on the effect of ceramic thickness on the dwell/penetration transition velocity and penetration velocity for commercially-available hot-pressed SiC in a baseline target configuration are presented and discussed.

EXPERIMENTAL SET-UP & PROCEDURES

A schematic illustration of the penetrator and target configurations used is shown in Figure 2. Monolithic SiC cylinders 38.1 mm in diameter and three different thicknesses, 12.7, 25.4, and 38.1 mm, were examined this study.

The commercially-available hot-pressed SiC used in this study was PAD SiC-N*. Table 1 lists some of the physical and mechanical properties determined for PAD SiC-N. Density and elastic moduli were determined using the Archimedes water immersion and pulse-echo techniques (ASTM Standard E494), respectively. Knoop hardness was determined on specimens final polished with 0.05 μm colloidal silica and in accordance with ASTM Standard C1326 using a load of 40 N. Fracture toughness was determined using the single-edge pre-cracked beam technique in accordance with ASTM Standard C1421 and specimen (bar) dimensions 3 mm x 4 mm x 50 mm. Flexural strength was determined from bars (30) of the same dimensions as those used for fracture toughness determination. A standard 20 x 40 mm semi-articulating flexure four-point fixture was used with a cross-head speed of 0.5 mm/min in accordance with ASTM C1161. Phases were determined using X-ray diffraction (Cu Kα, 0.02° 2θ step size, 2 sec dwell time), Orientation Imaging Microscopy, and commercial pattern matching programs. Grain size was determined using a commercial image analysis software package on scanning electron microscopy micrographs in which a minimum of 1000 grains were analyzed.

* Armor-grade pressure-assisted densification (PAD) SiC manufactured by BAE Advanced Ceramics, Vista, CA.

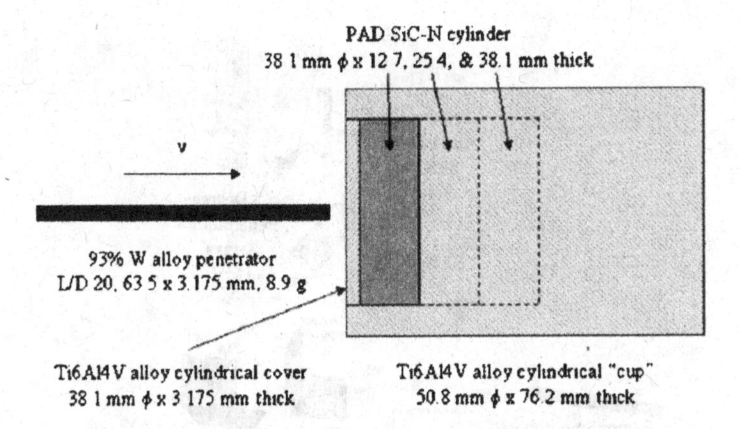

Figure 2. Penetrator and baseline target configuration used in this study.

Table 1. Physical and Mechanical Properties of PAD SiC-N

Density (g/cm^3)	Polytypes	Mean Grain Size (μm)	Young's Modulus (GPa)	Knoop Hardness (GPa)	Fracture Toughness (MPa*m$^{1/2}$)	Flexure Strength (MPa)
3.22	6H, 15R	1.90	452	18.6	5.1	620

The ceramic cylinders were slip-fitted into Ti6Al4V cups. The tight slip-fit allowed lateral confinement but no lateral pre-stress of the ceramic cylinders. The thick base of the Ti6Al4V cups was chosen to approximate a semi-infinite backing and allow for possible DOP measurements. Ti6Al4V cover plates 3.175 mm thick were welded to the cups.

The penetrators were W-based laboratory-scale L/D 20 rods with a length of 63.5 mm and a diameter of 3.175 mm. Rod density and mass were 17.6 g/cm^3 and 8.94 g, respectively. The penetrators were fired from a 27 mm smoothbore powder gun using launch packages which consisted of four-petal discarding sabot and a thin steel pusher with obturator. The launch package separates from the penetrator prior to impact on the ceramic target. Impact velocities were varied between 1000 and 1700 m/s.

Figure 3 is a schematic illustration of the instrumented forward-ballistic experimental set-up used in this study. Two sets of orthogonal striking 150 keV flash X-rays were used to determine the pitch/yaw and impact velocity of the penetrator. These X-rays were triggered indirectly by a break-screen through a digital time delay generator. Three separate 1 MeV flash X-rays (rotated 45° relative to one another) were used to capture the interaction between penetrator and ceramic target at specific times. A break-screen closer to the target was used for indirectly triggering the 1 MeV X-ray heads. Based on this set-up, the uncertainty in timing for the 1 MeV flash X-ray radiographs is estimated to be less than 1 μs.

The maximum uncertainty in the penetrator nose position measurement as determined from the flash X-ray radiographs is estimated to be less than one rod diameter. This large uncertainty is not due to image distortion due to parallax or misalignment of X-ray heads or the quality of the X-ray radiographs. Instead, it is due to poorly defined penetrator penetration heads, which often happens when penetration becomes non-steady. During non-steady penetration of a brittle material, penetrator material can flow into off-axis cone cracks or radial cracks ahead of the penetrator nose (rod center-

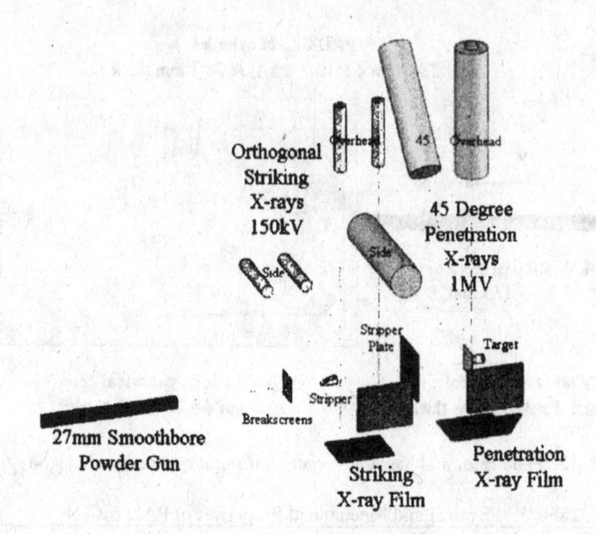

Figure 3. Schematic illustration of the experimental set-up showing the number and position of the 150 keV and 1 MeV flash X-rays.

line axis), masking its true penetration depth. Non-steady penetration would be expected for impact velocities near the dwell/penetration transition velocity or when the penetration velocity drops below that necessary for steady-state cavity formation.

EXPERIMENTAL RESULTS AND DISCUSSION

Flash X-ray Radiographs

Figure 4 shows a sequence of 1 MeV flash X-ray radiographs of a penetrator dwelling and subsequently, penetrating a 12.7 mm thick ceramic target. The impact velocity is 1216 m/s. The approximate location of the penetrator and ceramic are outlined in red and black, respectively. As can be seen, at 16 μs, the penetrator has fully penetrated the cover plate and is dwelling on the ceramic. The extent of lateral flow of penetrator material on the top of the ceramic indicates dwell has been going on for some amount of time. Furthermore, initial off-axis penetration of penetrator material into the ceramic can also be seen (white arrow, possible penetration into a cone crack). While this small amount of off-axis penetration of penetrator material does not mask the rod center-line location of the penetrator nose, it does demonstrate the potential error introduced as a result of the problem of being able to clearly identify the depth-of-penetration. At 38 μs, the penetrator has penetrated over half way through the ceramic, while at 59 μs, it has penetrated into the Ti6Al4V cup. Note in the 59 μs image, the cover plate has completely separated from the target.

Figure 5 is another sequence of radiographs showing interface defeat of the penetrator. The ceramic thickness is 38.1 mm and the impact velocity is 1207 m/s. At 49 μs, the penetrator has not penetrated into the ceramic; furthermore, the recovered ceramic did not show any penetration at the impact site.

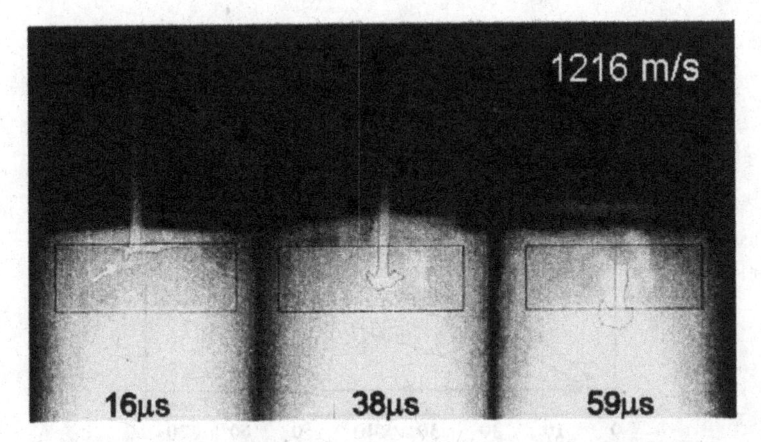

Figure 4. Sequence of three 1 MeV flash X-ray radiographs showing penetrator (red outline) dwelling followed by penetration of a 12.7 mm thick ceramic (black outline) target. v_{impact} = 1216 m/s.

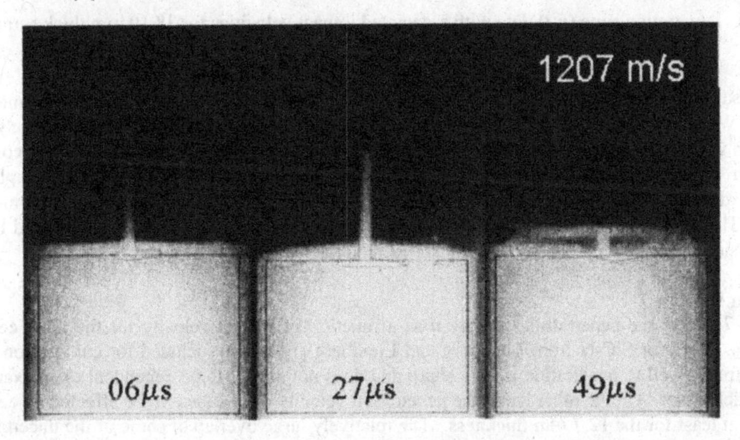

Figure 5. Sequence of three 1 MeV flash X-ray radiographs showing interface defeat of the penetrator (red outline) by a 38.1 mm thick ceramic (black outline) target. v_{impact} = 1207 m/s.

Penetration-Time (*p-t*) Data

Figure 6 shows *p-t* data determined from the 1 MeV flash X-ray radiographs for 38.1 mm thick ceramic target. The numbers next to the lines indicate the penetrator impact velocities. The horizontal lines demarcate the spatial positions of the interfaces between the cover plate and ceramic, and the ceramic and cup.

As can be seen in Figure 6, the change in velocity for the transition from interface defeat to penetration is relatively small, occurring between 1200 m/s and 1210 m/s. Therefore, the dwell/penetration transition velocity is estimated to be 1205 ± 5 m/s. This value was approximately the same for all three ceramic thicknesses.

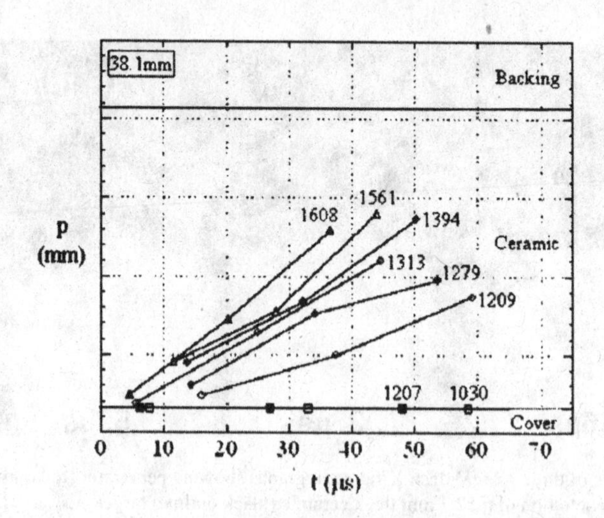

Figure 6. Penetration-time (*p-t*) data with associated impact velocities for 38.10 mm thick ceramic target.

It is also interesting to note that the *p-t* data essentially appear to be linear over the time intervals sampled. While it is acknowledged that additional data is needed to determine more precisely how penetration varies with time, it is not unreasonable to expect that steady-state penetration (i.e. constant penetration velocity) does occur for some period of time, especially at early times (negligible penetrator deceleration) for impact velocities much higher than the dwell/penetration transition velocity. This expectation is based on published experimental work [1-7] and also supported by the one-dimensional steady-state penetration equations of Alekseevski [13] and Tate [14].

Penetration Velocity

Figure 7 shows the penetration velocity *u* as a function of impact velocity for the three ceramic thicknesses. Data for SiC-N from Lundberg and Lundberg [1] is also included for comparison. The transition from dwell to penetration is very sharp and does not appear to be dependent upon examined ceramic thicknesses. On the other hand, the penetration velocity does appear to be affected by ceramic thickness, at least for the 12.7 mm thickness. The relatively large overlap of some of the uncertainties for the 25.4 mm and 38.1 mm thick ceramic targets possibly suggest that the two data sets are not distinct from each other. The uncertainties in the penetration velocities are predominately due to the uncertainties in position measurements of the penetrator nose and number of data points per impact velocity. In the following analysis, it is assumed that all data are distinct.

The dwell/penetration transition velocity and penetration velocities can be converted into relative ceramic penetration resistance values (R_T-Y_P) using the equations credited to Alekseevski [13] and Tate [14]. According to their analysis, the relative ceramic penetration resistance at the dwell/penetration transition is given by:

$$(R_T - Y_P)_{transition} = \frac{1}{2}\rho_P\left(v^{transition}\right)^2 \quad (1)$$

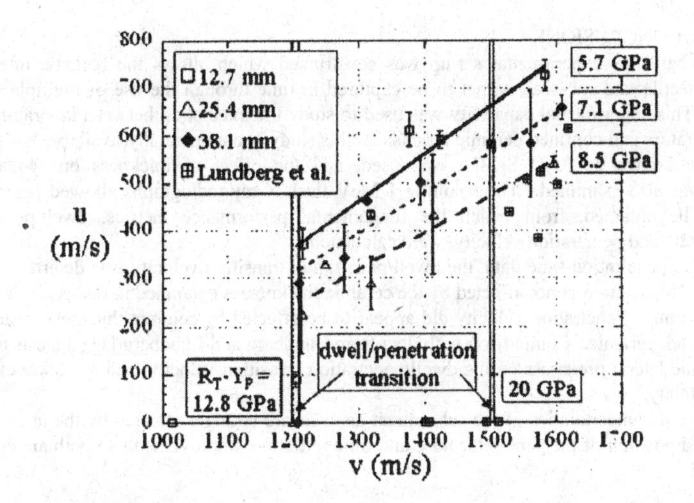

Figure 7. Penetration velocity (u) as a function of impact velocity for each ceramic thickness. Lundberg and Lundberg [1] data is also included for comparison.

where R_T is penetration resistance, Y_P is the effective dynamic yield strength of the penetrator (velocity-dependent value between the uni-axial stress and uni-axial strain values [15]), and ρ_P is the density of the penetrator. At the dwell/penetration transition, R_T-Y_P = 12.8 GPa. Using a slightly different target configuration, Lundberg and Lundberg [1] determined the dwell/penetration transition velocity to be 1507 ± 5 m/s for SiC-N, which corresponds to R_T-Y_P = 20 GPa. This is significantly higher than the value obtained in this study. The most significant difference between the two target configurations is that the ceramic cylinders in the Lundberg and Lundberg targets were laterally pre-stressed to 175 ± 25 MPa [1]. It will not be shown in this paper, but this is the primary reason for the higher ceramic penetration resistance in the Lundberg and Lundberg targets.

The relative ceramic penetration resistance value during penetration as determined from the Alekseevski [13] and Tate [14] equations is given by:

$$(R_t - Y_p)_{penetration} = \frac{1}{2}\rho_p v^2 \left(\frac{\rho_p}{\rho_p - \rho_t}\right)\left\{\left[1 - \left(1 - \frac{\rho_t}{\rho_p}\right)\left(\frac{u}{v}\right)\right]^2 - \frac{\rho_t}{\rho_p}\right\} \tag{2}$$

where v is the impact velocity and ρ_t is the density of the ceramic. Substitution of calculated penetration velocity values into Eqn. (2) and averaging (essentially assumed that the values are independent of impact velocity) yields relative ceramic penetration resistance values of 5.7 ± 0.7, 8.5 ± 0.8, and 7.1 ± 0.7 GPa during penetration for ceramic thicknesses 12.7, 25.4, and 38.1 mm, respectively. The value for the Lundberg and Lundberg [1] penetration velocity data is 9.5 ± 1.7 GPa. This value is not significantly higher than the values calculated for the 25.4 mm and 38.1 mm thick ceramics. This suggests that lateral pre-stress does not significantly enhance ceramic penetration resistance during penetration (at least for thick ceramics).

SUMMARY & CONCLUSIONS

A forward-ballistics experimental set-up was constructed which allows the ballistic interaction between a projectile and sub-scale target to be captured in time through the use of multiple 1 MeV flash X-rays. This experimental capability was used to study the interaction between laboratory-scale long-rod penetrators and confined ceramic targets. In this study, a commercially-available hot-pressed silicon carbide known as PAD SiC-N, was used, and the effect of thickness on fundamental performance was also examined. The resulting 1 MeV flash X-ray radiographs allowed penetration-time data to be obtained, from which the fundamental performance metrics, dwell/penetration transition velocity and penetration velocity, were calculated.

Based on the penetration-time data, the dwell/penetration transition velocity was determined to be 1205 ± 5 m/s. This value was not affected by the ceramic thicknesses examined in this study (i.e. 12.7, 25.4, and 38.1 mm). Penetration velocity did appear to be affected by ceramic thickness, at least for the 12.7 mm thick ceramic. Comparison with data from Lundberg and Lundberg [1] suggests a strong effect of ceramic lateral pre-stress on the dwell/penetration transition velocity, and a weaker effect on penetration velocity.

Lastly, since the uncertainties of the subsequent analysis are strongly affected by the uncertainties in the measured position of the penetrator nose and time initiation, improvements to both are currently being explored.

ACKNOWLEDGEMENTS

The authors wish to acknowledge the "unsung" ARL program managers, scientists, engineers, and technicians, both gone and those still with us, whose support made this study possible.

REFERENCES

1. P. Lundberg and B. Lundberg, "Transition between Interface Defeat and Penetration for Tungsten Projectiles and Four Silicon Carbide Materials," *Int. J. Impact Eng.*, **31,** 781-792 (2004).
2. D.L. Orphal and R. Franzen, "Penetration of Confined Silicon Carbide Targets by Tungsten Long Rods at Impact Velocities from 1.5 to 4.6 km/s," *Int. J. Impact Eng.*, **19**, 1-13 (1997).
3. D. L. Orphal, R. R. Franzen, A. C. Charters, T. L. Menna, and A. J. Piekutowski, "Penetration of Confined Boron Carbide Targets by Tungsten Long Rods at Impact Velocities from 1.5 to 5.0 km/s," *Int. J. Impact Eng.*, **19**, 15-29 (1997).
4. P. Lundberg, R. Renstrom, and B. Lundberg, "Impact of Metallic Projectiles on Ceramic Targets: Transition between Interface Defeat and Penetration," *Int. J. Impact Eng.*, **24**, 259-275 (2000).
5. P. Lundberg, R. Renstrom, and B. Lundberg, "Impact of Conical Tungsten Projectiles on Flat Silicon Carbide Targets: Transition from Interface Defeat to Penetration," *Int. J. Impact Eng.*, **32**, 1842-1856 (2005).
6. T. Behner, D.L. Orphal, V. Hohler, C.E. Anderson, Jr., R.L. Mason, and D.W. Templeton, "Hypervelocity Penetration of Gold Rods into SiC-N for Impact Velocities from 2.0 to 6.2 km/s," *Int. J. Impact Eng.*, **33**, 68-79 (2006).
7. M.W. Burkett, R.B. Parker, A.D. Rollett, and G.E. Cort, "FY 90 PHERMEX and Q-Site Confined Ceramic Penetration Experiments (U)," Los Alamos Technical Report, LA-12099-MS, August 1991, 156 pp.
8. G.E. Hauver, E.J. Rapacki, Jr., P.H. Netherwood, and R.F. Benck, "Interface Defeat of Long-Rod Projectiles by Ceramic Armor", ARL Technical Report, ARL-TR-3590, September 2005, 85 pp.
9. D. Sherman, "Impact Failure Mechanisms in Alumina Tiles on Finite Thickness Support and the Effect of Confinement," *Int. J. Impact Eng.*, 24, 313-328 (2000).
10. D. Sherman and D.G. Brandon, "The Ballistic Failure Mechanisms and Sequence in Semi-

Infinite Supported Alumina Tiles, *J. Mat. Res.*, **12**, 1335-1343 (1997).

11. I.M. Pickup, A.K. Barker, I.D. Elgy, G.J.J.M. Peskes, and M. van de Voorde, "The Effect of Coverplates on the Dwell Characteristics of Silicon Carbide Subject to KE Impact," Proceedings of the 21st International Symposium on Ballistics, Adelaide, Australia (2004).
12. J.C. LaSalvia, E.J. Horwath, E.J. Rapacki, C.J. Shih, and M.A. Meyers, "Microstructural and Micromechanical Aspects of Ceramic/Long-Rod Projectile Interactions: Dwell/Penetration Transitions," Fundamental Issues and Applications of Shock-Wave and High-Strain-Rate Phenomena, eds. K.P. Staudhammer, L.E. Murr, and M.A. Meyers, Elsevier Science, 2001, 437-446.
13. V.P. Alekseevski, "Penetration of a Rod into a Target at High Velocity," *Fiz. Goren. Vzryra*, **2**, 99 (1966).
14. A. Tate, "A Theory for the Deceleration of Long Rods After Impact," *J. Mech. Phys. Solids*, **15**, 387 (1967).
15. V. Hohler and A.J. Stilp, "Long-Rod Penetration Mechanics," High Velocity Impact Dynamics, ed. J.A. Zukas, John Wiley 7 Sons, Inc., 1990, New York, 321-404.

INSTRUMENTED HERTZIAN INDENTATION STUDY OF TWO COMMERICAL SILICON CARBIDES

H.T. Miller*, R.C. McCuiston**, J.C. LaSalvia
U.S. Army Research Laboratory
AMSRD-ARL-WM-MD
Aberdeen Proving Ground, MD 21005-5069

ABSTRACT

Of the armor ceramics commercially available today, hot-pressed silicon carbides (SiCs) are very effective against a wide range of ballistic threats. Pressureless-sintered SiCs are attractive alternatives to hot-pressed variants for some particular applications based upon cost. The microstructural and mechanical property differences between hot-pressed and pressureless-sintered variants result in different damage mechanisms and/or damage kinetics which are qualitatively believed responsible for differences in their ballistic performances. Characterizing and quantifying the differences in microstructure, mechanical properties, and damage behavior is therefore important for building the connections between processing, microstructure, properties, and performance. Two commercially-available SiC variants, Saint Gobain Hexoloy (pressureless-sintered) and BAE Systems SiC-N (hot-pressed), were characterized and examined using instrumented Hertzian indentation in order to quantitatively and qualitatively compare their damage responses to very large localized stresses. Spherical diamond indenters with nominal diameters of 0.5 mm and 5.0 mm were used to explore both a quasi-plastic mode response and brittle mode response in these ceramics. Acoustic emission data was acquired during the indentation testing in an attempt to correlate loads with specific macrocrack-types. Optical and electron micrographs of the samples will be shown.

INTRODUCTION

In an effort to design better materials for existing and future applications, researchers are currently striving for new and innovative ways to unlock critical information into the potential performance of materials. Microstructural characteristics can attribute to different damage mechanisms and damage kinetics in armor-grade ceramics that consequently have a significant effect on performance. Hertzian instrumented indentation is a simple and useful tool in qualitatively and quantitatively evaluating damage mechanisms in ceramic materials.[1] The benefit of using Hertzian instrumented indentation is that it tests the elastic and plastic response of the material, unlike completely plastic indentation techniques such as Vickers and Knoop.[2] Thus, this technique allows for a full response history to be evaluated, and hopefully insight into the relation of microstructure and damage behavior on performance.

Two primary damage modes can occur when indenting a ceramic with a spherical indenter. "Brittle" mode fracture, characterized by the formation of Hertzian cone cracks, is a tensile driven failure initiated by a surface flaw. "Quasi-plastic" mode fracture is a shear-driven compressive failure mechanism consisting of shear faults or micro-cracking. Microstructure and indenter diameter are factors that contribute to the damage modes initiated and hence the lifetime limiting factors of the material.[3-4]

Figure 1 shows representative microstructures of SiC-N and Hexoloy. The presence of an aluminum and oxygen rich intergranular glassy phase in hot-pressed SiC-N would be expected to experience a more "quasi-plastic" response when spherically indented.[5] Furthermore, Hexoloy should fail in a more "brittle" mode due to lack of an extensive intergranular glassy phase and the presence of porosity and large graphitic particles.[6] As can be seen, abnormally large grains, and an increase in

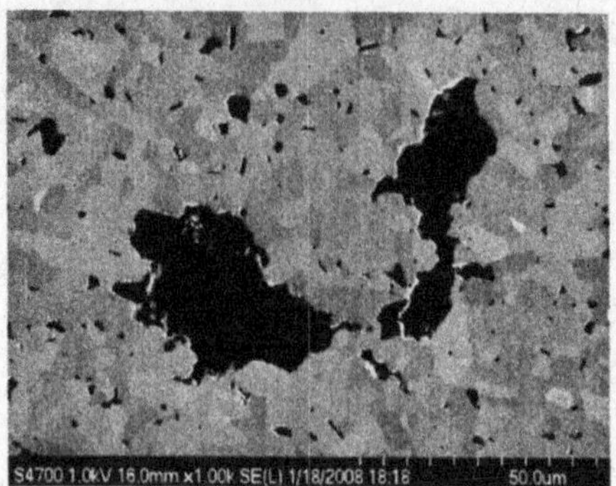

Figure 1. Representative microstructures of pressure-less sintered Hexoloy (left) and hot-pressed SiC-N (right).

porosity and excess carbon were found in the Hexoloy material that was tested. whereas. SiC-N has a smaller grain size, and no apparent porosity.

EXPERIMENTAL PROCEDURE

A pressureless-sintered silicon carbide (Hexoloy, Saint-Gobain, France), and a hot-pressed silicon carbide (SiC-N, BAE Systems, Vista CA) were used for the indentation experiments. Samples of approximately 50 mm x 25 mm x 12 mm were mounted and polished using standard metallographic procedures to a finish of 1 µm. The polished samples were indented with spherical diamond indenters having nominal diameters of 0.5 mm and 5.0 mm. The indenters were manufactured using a Precision Spindle Method™ (Gilmore Diamond Tools. Attleboro MA) that resulted in a indenter tip with minimal surface flaws.[4] A material testing machine (Z005, Zwick/Roell, Atlanta GA) having a 2500 N load cell capacity was used to generate the indents. The loading/unloading rate of the cross-head was a constant 1 µm/s. A series of indents were made at loads of 5 N – 200 N with the 0.5 mm diameter indenter, and 100 N – 2000 N with the 5.0 mm diameter indenter.

Upon indenting the materials, acoustic emission measurements were continuously recorded throughout the tests. An acoustic emission sensor (Physical Acoustics Corporation, Princeton Junction NJ) was affixed to the indenter, and emission data was recorded using Physical Acoustics' AEWin software. The acoustic emission activity was parametrically combined with the load cell data of the material testing machine to allow the acoustic emission data to be graphically related to the corresponding load. An amplitude of 35 dB was determined to be a suitable threshold level for distinguishing acoustic emission events. Indentation surfaces were examined optically using differential interference contrast (Nomarski), as well as using a scanning electron microscope (S4700, Hitachi, Pleasanton CA).

RESULTS AND DISCUSSION

5 mm Diamond Indenter Results

Optical images of the initial observable indents (5.0 mm diamond indenter) of Hexoloy and SiC-N and their corresponding acoustic emission outputs are shown in Figure 2. The initial observable indent in Hexoloy occurred at 300 N (6.3 GPa) and in SiC-N at 800 N (8.8 GPa). The mean elastic contact pressure[7] of these loads are in parenthesis, and are only valid until the yield point of the

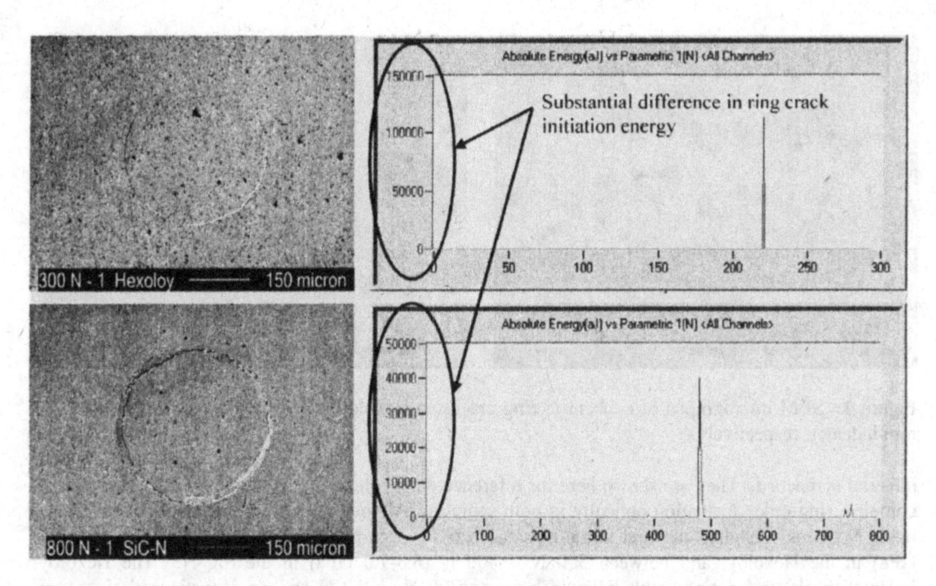

Figure 2. Nomarski images of initial observable indents in Hexoloy and SiC-N using a nominal 5.0 mm spherical diamond indenter and their corresponding acoustic emission signals.

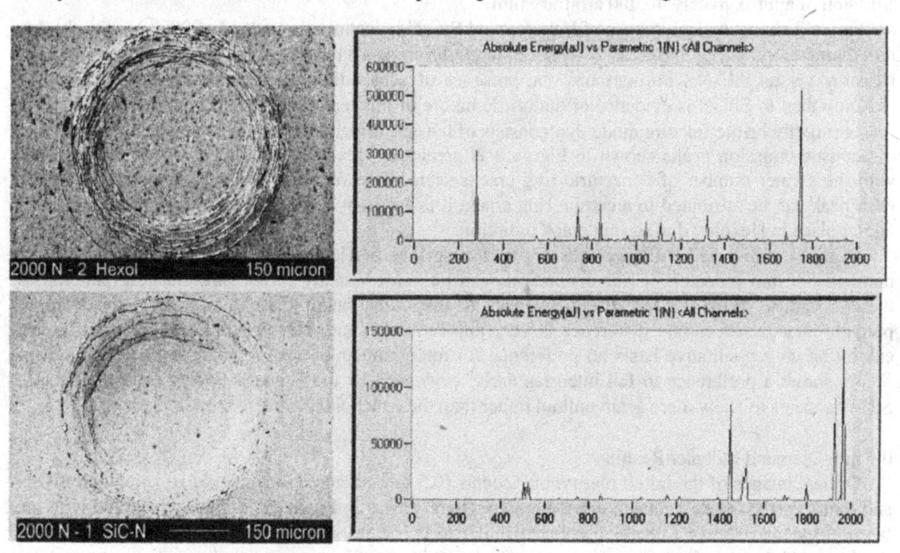

Figure 3. Nomarski images of Hertzian indentations in Hexoloy and SiC-N at 2000 N using a nominal 5.0 mm spherical diamond indenter and their corresponding acoustic emission signals.

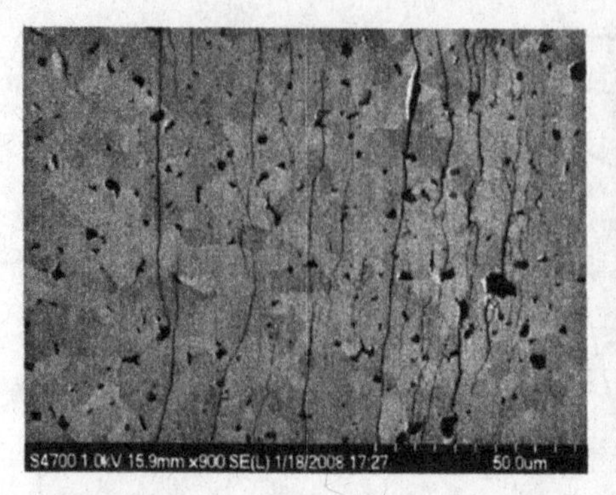

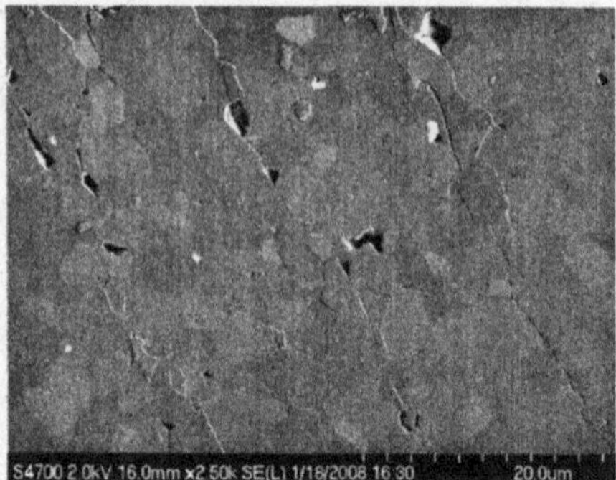

Figure 4. SEM micrographs of concentric ring cracks in Hexoloy and SiC-N using the nominal 5.0 mm indenter, respectively.

material is reached. They are shown here for reference only. These images show the first evidence of complete ring crack formation optically in both samples. Although these are the first visible indents using Nomarski imaging, the first initial acoustic activity occurred between 100 N – 200 N (4.4-5.5 GPa) in the Hexoloy, and between 300 N – 400 N (6.3-7.0 GPa) in the SiC-N. The Hexoloy indentation showed a ring crack initiated between 200 N – 250 N and an acoustic emission over 100,000 arbitrary units. SiC-N experienced a ring crack initiation at nearly 500 N and an acoustic emission of approximately 40,000 arbitrary units.

Figure 3 shows optical images of Hexoloy and SiC-N indentations at a load of 2000 N with the 5.0 mm diameter indenter. Similarly to Figure 2, large differences in acoustic emission energy are seen in Hexoloy versus SiC-N. Furthermore, the presence of substantially more concentric ring cracks in Hexoloy than in SiC-N is evidence of the brittle nature of this material.[2] The 5.0 mm indenter allows for a primarily brittle fracture mode that consists of tension-driven Hertzian cone cracks.[8] The number of acoustic emission peaks shown in Figure 3 is greater in Hexoloy than SiC-N which is consistent with the greater number of concentric ring cracks seen in Hexoloy. While it is not known whether each peak can be attributed to a certain ring crack, it is apparent that a greater strain energy release is taking place in Hexoloy during ring crack initiation.

Figure 4 shows SEM micrographs of the top surface of Hexoloy and SiC-N indents using the nominal 5.0 mm indenter. Clear distinctions can be made visually in the ring cracking patterns of these ceramics. First, despite the differences in magnification of these two images, the level of porosity and excess carbon difference is very apparent. Second, Hexoloy has straight crack paths, exhibiting on a qualitative basis no preference for intergranular or transgranular fracture. However, SiC-N shows a preference to fail intergranularly, evidenced by the jagged cracking pattern. Finally, SiC-N appears to show more grain pullout rather than the conchoidal fracture seen in Hexoloy.

0.5 mm Diamond Indenter Results

Optical images of the initial observable indents (0.5 mm diamond indenter) of Hexoloy and SiC-N and their corresponding acoustic emission outputs are shown in Figure 5. Hexoloy indented with the 0.5 mm indenter shows a cracking pattern similar to its indent with the 5.0 mm indenter. That is, it displays a higher level of acoustic activity than SiC-N, the frequency of the acoustic emissions are greater, and the indent does not show evidence of residual depth when viewed optically. The indent in SiC-N appears to be qualitatively different. The residual depth of the SiC-N sample shown in

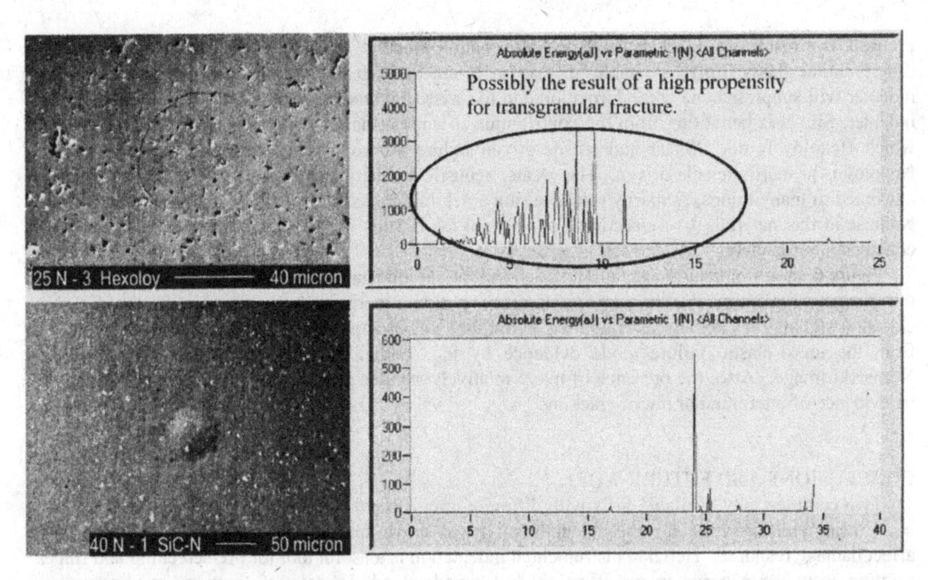

Figure 5. Nomarski images of initial observable indents in Hexoloy and SiC-N using a nominal 0.5 mm spherical diamond indenter and their corresponding acoustic emission signals.

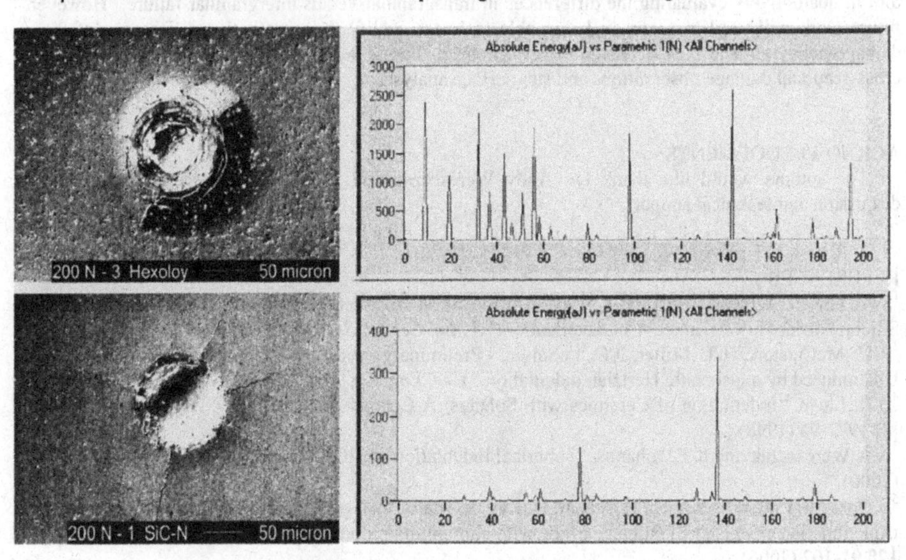

Figure 6. Nomarski images of Hertzian indentations in Hexoloy and SiC-N at 200 N using a nominal 0.5 mm spherical diamond indenter and their corresponding acoustic emission signals.

Figure 5 is a result of a local shear-driven deformation process.[9] This is a result of a weak interface (intergranular glassy film) that enables diffuse micro-cracking. It is well known that using a "sharper" indenter will suppress cone crack formation and likewise ring crack formation.[10-11] Using the 0.5 mm indenter, SiC-N is benefiting from the contribution of another damage mode (quasi-plastic failure) in which Hexoloy is not. Shear and tensile-driven failure are competing responses in SiC-N, while Hexoloy is primarily tensile-driven. The greater acoustic intensity and frequency in Hexoloy could be attributed to many things, primarily multiple ring crack initiations, sub-surface macrocracking, or pore collapse in the material. Cross-sectional analysis of the indents would be needed to complement this conclusion completely.

Figure 6 shows optical images of Hexoloy and SiC-N indentations at a load of 200 N with the 0.5 mm diameter indenter. Again, a large difference in strain energy release is visible via the acoustic emission signals for Hexoloy as compared to SiC-N. SiC-N appears to show a stronger contribution from the quasi-plastic failure mode evidence by the residual depth displacement shown in the Nomarski image. Also, the presence of many relatively smaller peak acoustic emission signals could be evidence of intergranular micro-cracking.

CONCLUSIONS AND FUTURE WORK

Two commercially available SiCs with different microstructural characteristics were mechanically tested using Hertzian instrumented indentation. It was shown that microstructural characteristics can affect damage response. Hertzian instrumented indentation is a useful tool for pre-screening and finger printing response patterns in an effort to better understand the role of fracture, microstructure, processing, and mechanical behavior on performance. Acoustic emission was shown to be a useful tool in qualitatively evaluating the differences in transgranular versus intergranular failure. However, future work will need to focus on being able to better qualify the quantitative differences of these damage behaviors and how it relates to performance. Future work will also be progressing towards cross-sectional damage observations and sub-surface analysis.

ACKNOWLEDGEMENTS

The authors would like thank Dr. Andy Wereszczak and Mr. Russell Kress for their helpful discussion and technical support.

REFERENCES

[1]S.K. Lee, S. Wuttiphan, and B.R. Lawn, "Role of Microstructure in Hertzian Contact Damage in Silicon Nitride: I. Mechanical Characterization," *J. Am. Ceram. Soc.*, **80** [9] 2367-81 (1997).

[2]R.C. McCuiston, H.T. Miller, J.C. LaSalvia, "Preliminary investigation of damage in armor-grade B_4C induced by quasi-static Hertzian indentation," *Cer. Eng. Sci. Proc.*, **28**(5), 171-80 (2007).

[3]B.R. Lawn, "Indentation of Ceramics with Spheres: A Century after Hertz," *J. Am. Ceram. Soc.*, **81** [8] 1977-94 (1998).

[4]A.A Wereszczak and K.E. Johanns, "Spherical Indentation of SiC," *Cer. Eng. Sci. Proc.*, **27**(7), (2006).

[5]E. Pabit, S. Crane, K. Seiben, D.P. Butt, D. Ray, R. Marc Binders, R.A. Cutler, "Grain boundary and triple junction chemistry of silicon carbide with aluminum or aluminum nitride additive," *Cer. Trans.*, **178** 91-102 (2006).

[6]W.J. Moberlychan, J.J. Cao, L.C. de Jonghe, "The roles of amorphous grain boundaries and the β-α transformation in toughening SiC," *Acta mater.*, **46** [5] 1625-35 (1998).

[7]K.L. Johnson, Contact Mechanics, Cambridge University Press, New York, 1985.

[8]Y. Rhee, H. Kim, Y. Deng, and B.R. Lawn, "Brittle Fracture versus Quasi Plasticity in Ceramics: A Simple Predictive Index," *J. Am. Ceram. Soc.*, **84** [3] 561-65 (2001).
[9]B.R. Lawn, N.P. Padture, H. Cai, and F. Guiberteau, "Making Ceramics "Ductile"," *Science (Washington, D.C.)*, **263**, 1114-16 (1994).
[10]M.V. Swain and J.T. Hagan, "Indentation Plasticity and the Ensuing Fracture of Glass," *J. Phys. D: Appl. Phys.*, **9**, 2201-14 (1976).
[11]K.E. Puttick, "Energy Scaling, Size Effects, and Ductile-Brittle Transitions in Fracture," *J. Phys. D: Appl. Phys.*, **12**, L19-L23 (1979).

APPARENT YIELD STRENGTH OF HOT-PRESSED SiCs

W. L. Daloz and A. A. Wereszczak
Ceramic Science and Technology
Oak Ridge National Laboratory
Oak Ridge, TN 37831

O. M. Jadaan
College of Engineering, Mathematics, and Science
University of Wisconsin-Platteville
Platteville, WI 53818

ABSTRACT

Apparent yield strengths (Y_{App}) of four hot-pressed silicon carbides (SiC-B, SiC-N, SiC-HPN, and SiC-SC-1RN) were estimated using diamond spherical or Hertzian indentation and the considerations of the von Mises and Tresca yield criteria. The developed test method was robust, simple and quick to execute, enabling the acquisition of confident sampling statistics. The choice of indenter size, test method, and method of analysis are described. The compressive force necessary to initiate apparent yielding was identified postmortem using differential interference contrast (or Nomarski) imaging with an optical microscope. It was found that the Y_{App} of SiC-HPN (14.0 GPa) was approximately 10% higher than the equivalently valued Y_{App} of SiC-B, SiC-N, and SiC-SC-1RN. This distinction in Y_{App} shows that the use of this test method could be valuable because there were no differences among the average Knoop hardnesses of the four SiC grades.

INTRODUCTION

When a contact force is applied by an indenter, large compressive forces can arise that create a plastic-like deformation in a ceramic. While similarity can exist in appearance to plastic deformation in metals, no "classical" plastic flow occurs so this response in ceramics can be validly termed "apparent yielding". Nomenclature aside, the apparent yield strength (Y_{App}) of a ceramic is informative because it can be used to assess and rank contact-damage resistance in ceramics where such a characteristic is desired, for example, ceramics for wear applications and armor ceramics. Additionally, its value can be used as one of many required inputs for computer models to predict ballistic response of ceramic armor[1].

Because ceramic materials tend to be very brittle, and have low tensile strength concomitant with very high compressive strength, traditional methods of measuring yield strengths (e.g., uniaxial tensile testing) do not work. Ceramics undergo brittle fracture in tension at much lower forces than at the loads or stresses where apparent yielding initiates in them.

Indentation testing provides a relatively cheap, quick and simple alternative that can be executed using fairly common and inexpensive equipment and requires minimal specimen preparation. Because of the small size of the indents (e.g., a diameter of many tens or hundreds of microns) relative to the rest of the sample, the distorted field is well confined by an elastic semi-solid and much higher compressive stresses can accrue before critical tensile stresses are reached that initiate cracking. Furthermore, even with a small sample area less than 5 cm^2, enough tests can be run to obtain a reliable and reproducible value of Y_{App}, and the testing and analysis can be completed in a matter of days rather than weeks or months.

Spherical indentation can produce damage similar to a ballistic impact, and unlike angled, square or diamond shaped indenters, the elastic limit or Y_{App} can be determined because a finite load can be applied where there is only an elastic response. The evolution of failure mechanisms during Hertzian indentation is illustrated in Fig. 1 for the use of a "large" or "small" diameter indenter. Ring cracking is more apt to initiate before apparent yielding as indenter diameter is increased and the reverse is more likely to occur as indenter diameter decreases. This "sharp" to "blunt" indenter effect on target material has been described by Mouginot[2]. In order to quantify the apparent yield behavior, it is important that yielding be the only active mode of failure so that it is not complicated by the pre-existence or concurrence of fracture. This can be accomplished with the use of a sufficiently small diameter indenter.

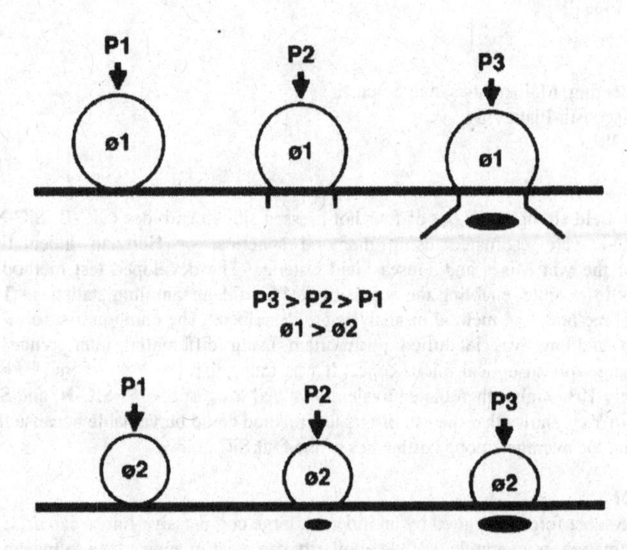

Figure 1. Schematic of contact damage evolution during loading and unloading of a Hertzian indentation test for "large" indenter diameters (top). Using a "small" diameter indenter (bottom) causes apparent yielding to initiate before ring crack initiation (i.e., sought in this study).

Additionally, indenter-related considerations can arise with Hertzian indentation of hard ceramics. WC/Co cermet indenters, whose use tend to dominate the spherical indentation literature, are not as hard as many ceramics such as B_4C and many SiCs. Thus, use of these can introduce indenter-deformation complications into the interpretation of the target's response. In order to ensure that an indenter does not permanently deform during testing and unexpectedly alter the generated stress field in the target ceramic, the use of a diamond indenter is preferred here for investigating yield-like responses of hard ceramics. In addition to diamond's high hardness and resistance to permanent deformation, its very high elastic modulus (~1140 GPa) is also valuable for its ability to promote yielding before brittle failure. Work by Johnson[3], which the authors have confirmed[4], shows that mismatch between the elastic moduli of the target and indenter causes friction at their contact area.

When the indenter's elastic modulus is greater than the target's, the Poisson effect differences between the two cause friction that in turn creates a radial compressive stress that counters the surface radial tensile stress that is just outside the contact area - a stress that causes ring cracking. This allows greater subsurface compressive stresses to be reached before ring cracks form, and therefore, a high elastic modulus indenter like diamond is more likely to promote yielding as the failure mechanism to initiate first. Because the very high compressively-confined shear stresses that initiate apparent yielding form below the surface at approximately a depth of one-half the contact radius (Fig. 6), the effects of friction at the surface have negligible effect on the value of maximum Y_{App}[5].

PROCEDURE

The indentation testing was done with a computer-controlled instrumented indentation system (MTIC, Oak Ridge, TN). A computer controls the X and Y motion of a stage between an optical microscope and the indenter. The spherical indenter is at the bottom of the Z stage or load train and is translated at a set displacement rate with a computer-controlled stepper motor. Load is measured with a load cell and recorded by the computer. A camera, positioned toward the point of contact, is used to observe and guide the coarse approach of the indenter as it is lowered onto the target. Samples are first mounted on the stage and aligned under the microscope. Once the user has set the X and Y coordinates of the indents, the stage then automatically translates into position under the indenter. The indenter moves downward at a user pre-determined controlled displacement rate (0.1 μm/s) until the desired maximum load is reached. Displacement controlled unloading follows this.

A custom 500-μm-diameter diamond indenter (Gilmore Diamond Tool, Attleboro, MA) shown in Fig. 2 was used for the testing. This size enables a large compressive stress to be produced (i.e., those that initiate yielding) at a modest compressive force prolonging the life of the expensive indenter. As previously discussed, larger diameter indenters are more likely to initiate ring cracks before apparent yielding whereas smaller diameter indenters are more likely to initiate apparent yielding first. With further decreasing indenter size however, stresses can become very high even with small compressive forces. Confident identification of the apparent yield initiation is therefore difficult and often subjective to achieve. Thus, a 500 μm diameter size is an effective compromise for studying yielding behavior.

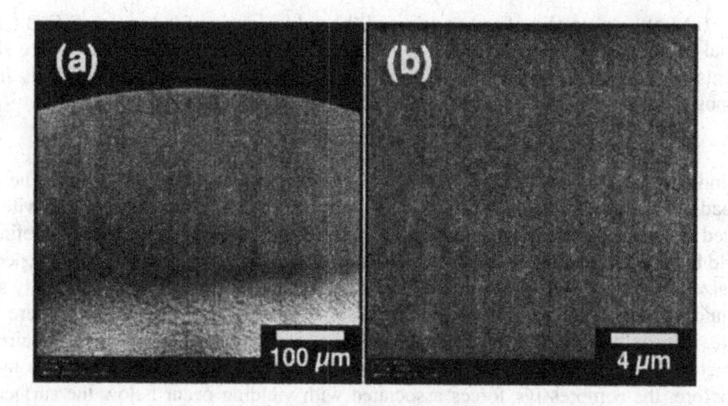

Figure 2. (a) SEM image of indenter tip (b) close-up of polished region.

The test algorithm is illustrated in Fig. 3. An indent test matrix was set up as shown in Fig. 3(a). The specimen surface was first cleaned with acetone in order to remove any debris or oils that could interfere with indentation or mask identification of apparent yield initiation. Five indents were spaced 0.5 mm apart with increasing load increments of 4.5 N up to a maximum load between 22 to 40 N. The indents were then examined postmortem for indications of apparent yielding. The highest load without apparent yielding and the lowest load with apparent yielding were used to estimate the load at which apparent yielding had initiated. Once a load range was bracketed to +/ 4.5 N, another fifteen sets of five indents each were generated, this time in increments of 2.25 N about the previously determined yield range. For example, if the first test showed yielding at 31 N and no yielding at 26.5 N, then the following tests would be run varying loads from 24.5 N to 33 N at 2.25 N increments. Such finer bracketing resulted in better resolution of the identified yield initiation forces.

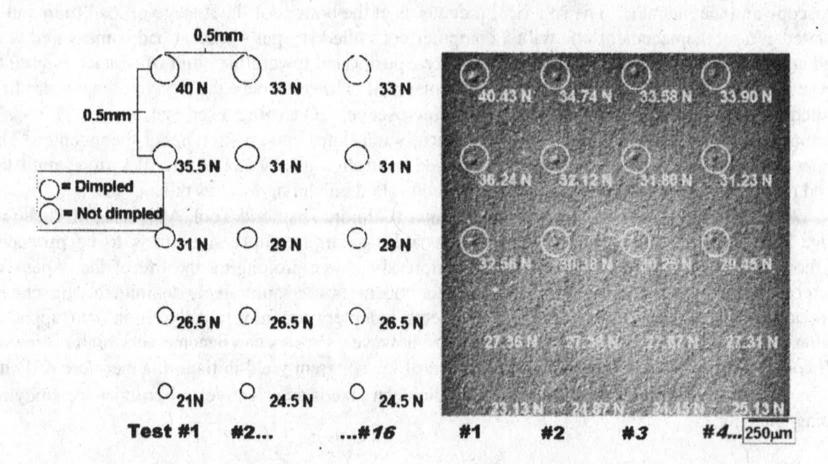

Figure 3. (a) Left - schematic of test matrix and (b) right - image of indents on SiC-B taken in differential interference contrast (Nomarski imaging). Actual compressive forces are shown. The indents that had yielded are circled with a bold circle and the dimples are evident. In each test it is possible to identify a load below which yielding was not observed.

The indent load data were continuously recorded on the test system's computer and the actual maximum loads were extracted. The highest load without dimpling and the lowest load with dimpling were averaged for each of the sixteen tests, and the average of those averages was defined as the apparent yield load for the sample. Three samples of each material were tested to verify repeatability.

The size of the sample only needs to be large enough to accommodate a statistically significant number of indents without them overlapping or interfering, and in this case, a 2 cm square was more than adequate. This is attractive because a material's Y_{App} can be determined without requiring a large amount of material. The surfaces were polished to a ¼-micron finish using diamond polishing. As mentioned before, the compressive forces associated with yielding occur below the surface, and are only minimally affected by surface condition. However, a smooth polished surface makes it easier to optically identify when yielding has occurred which is crucial to this method.

The onset of yielding was determined by direct microscopic observation of the indents in differential interference contrast (DIC or Nomarski) imaging using a standard metallographic microscope (Olympus Model BX51M, Orangeburg, NY). With a clean and smooth surface, DIC microscopy allows straightforward imaging of the shallow residual dimples left when yielding has occurred, as shown in Fig. 3(b). Initially, capacitance gages were used to continuously measure indenter displacement during indentation testing enabling the construction of a depth of penetration vs. load curve for each test. Hysteresis can be seen between the load and unload curves when yielding occurs[6] but was often difficult to confidently discern when the amount of yielding was very small. This method proved inconsistent in identifying the onset of yielding, as loads below 45 N sometimes showed no hysteresis but under a microscope showed apparent yielding had initiated. Subjectivity needed to be avoided, so optical observation with DIC imaging was relied upon with much more confidence because the dimpling caused by indentation was always distinct, as can be seen in Fig. 3(b) where dimples at each indent location were clearly either present or missing.

The materials tested in this study were four different hot pressed silicon carbide ceramic materials, SiC–B, SiC–N, SiC–SC–1RN and SiC–HPN (all fabricated by BAE Advanced Ceramics, Vista, CA). Hot pressing allows the materials to be formed to 100% of their theoretical density which is desirable for many contact-related applications because porosity can decrease the initiation force for apparent yielding. All four of these SiCs show an indentation size effect in their Knoop hardnesses, but more notably, their average Knoop hardnesses are statistically equivalent as shown in Fig. 4. Their elastic properties were measured using resonant ultrasound spectroscopy using a method described elsewhere[7] and their values are listed in Table I. The microstructures of the four SiCs are shown in Fig. 5. The SiC-B and SiC-N have a slightly coarser average grain size than the SiC-SC-1RN and SiC-HPN.

Table I. Elastic properties of the four SiC grades measured using resonance ultrasound spectroscopy.

Grade	Elastic Modulus (GPa)	Poisson's Ratio
SiC-B	446	0.154
SiC-N	453	0.171
SiC-HPN	443	0.155
SiC-SC-1RN	431	0.154

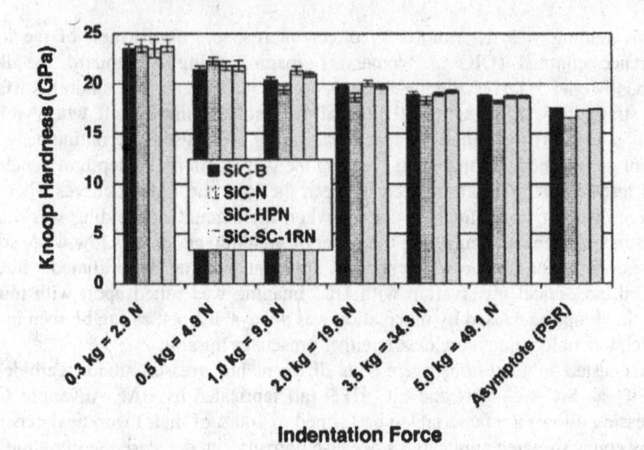

Figure 4. Average Knoop hardness of SiC-B, SiC-N, SiC-HPN and SiC-SC-1RN as a function of indentation force. Error bars show standard deviation. The "asymptotic" value represents a calculated load independent hardness according to the proportional specimen resistance (PSR) model (Li & Bradt)[8]. There are no significant differences in the average Knoop hardnesses among the four SiCs.

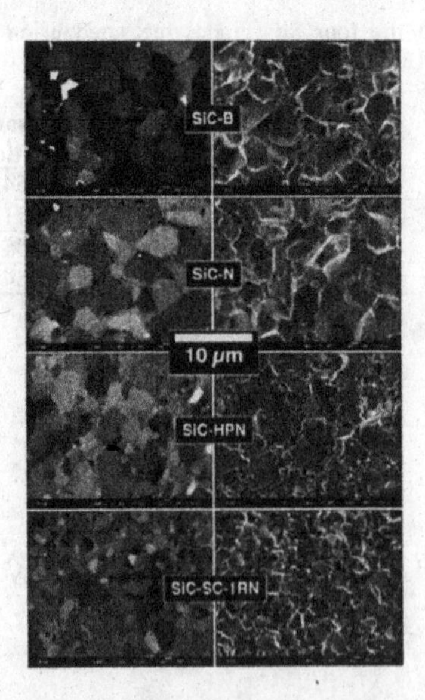

Figure 5. Microstructures of the four SiCs; polished left, fracture surfaces right.

YIELD STRESS CRITERIA

Ceramics typically fail from critically large first principal tensile stresses. But because the damage within the target under the indenter is plastic-like and not brittle, the Tresca and von Mises yield criteria are more appropriate for representing the apparent yielding. The von Mises criterion considers that yielding will occur when the distortion energy per unit volume in a component under a multiaxial stress state becomes equal or greater than the distortion energy per unit volume at the time of yielding in a simple uniaxial stress test using a specimen of the same material.[5,9] The von Mises yield stress criterion is then given by:

$$\frac{1}{\sqrt{2}}\left[(s_1 - s_2)^2 + (s_2 - s_3)^2 + (s_3 - s_1)^2\right]^{1/2} \geq Y \;. \qquad (1)$$

The von Mises yield criterion is representative of most metals and as such is widely available in modeling software.

In ceramics however, plastic-like deformation does not occur by the same classic flow mechanisms, but rather by shear driven processes[10]. As an alternative, the Tresca criterion considers maximum shear stress as the causal factor in yielding, rather than distortion energy as in von Mises. For the Tresca criterion, yielding is predicted to occur when the maximum shear stress in a component under multiaxial stress state becomes equal or greater than the maximum shear stress at the time of yielding in a simple uniaxial stress test using a specimen of the same material. Hence, yielding is predicted to occur if any of the following expressions occur:

$$\begin{aligned} |s_1 - s_2| &\geq |Y| \\ |s_2 - s_3| &\geq |Y| \;. \\ |s_3 - s_1| &\geq |Y| \end{aligned} \qquad (2)$$

Under Hertzian indentation, the maximum shear stress occurs along the contact axis (line of loading) at a depth of approximately $0.5a$, where a is the radius of the contact area. Along this contact axis of symmetry, $s_1 = s_2$. Substituting this condition in Eqs. 1 and 2 shows that for the case of Hertzian indentation, the two yield criteria become equivalent up to the onset of yielding and predict yielding to occur when:

$$|s_3 - s_1| \geq |Y| \;. \qquad (3)$$

When applied to Hertzian indentation the stress to initiate yielding is related to the maximum contact pressure, giving rise to the equation:[10,11]

$$P_Y/r^2 = (1.1\pi Y)^3 (4k/3E)^2 = const \;, \qquad (4)$$

where P_Y is the force where apparent yielding initiated, E is elastic modulus of the target, and r is the radius of the indenter, and $k = (9/16)\left[(1-v^2)+(1-v^{*2})E/E^*\right]$ where v is Poisson's ratio and * indicates the indenter material.

Figure 6 shows the von Mises and Tresca stress fields created by the Hertzian loading of SiC with a 500-μm-diamond indenter when the SiC is purely elastic. As expected, the likeness of the images confirms that the two criteria are equivalent in this case. The same is true of the plastic case when yielding occurs.

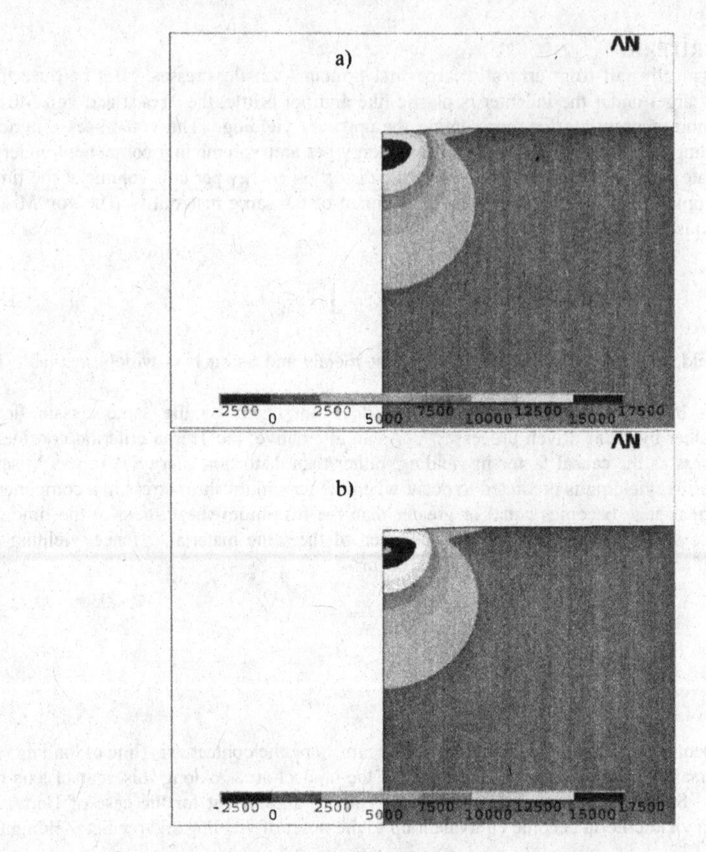

Figure 6. FEA model of (a) the von Mises and (b) the Tresca stress fields created during 50 N Hertzian loading of SiC (E=450 GPa and ν=0.16) with a 500 μm diamond indenter. The SiC is elastic in this example. Legend units are in MPa.

Figure 7 shows the von Mises stress field when the target is (a) purely elastic, and (b) when its yield strength is taken to be 13 GPa. The indentation creates very high compressive and shear stresses directly under the center of the indenter and below the target surface. The maximum surface stress occurs directly under the indenter at the center of the contact and is 50% greater than the average contact stress. Figure 7(c) shows the distribution of the first principal stress for comparison. The maximum first principal tensile stress occurs at a ring location surrounding the contact area. These radial tensile stresses can cause brittle ring crack failure. They must be minimized for apparent yield testing in order to promote quasi-plasticity as the dominant mode of failure, or, the latter must be promoted somehow at the expense of the former (e.g., use a small diameter indenter).

RESULTS AND DISCUSSION

All four SiCs exhibited distinct quasi-plastic deformation. All produced identifiable dimples for loads above 40 N without ring cracking. This deformation is similar in appearance to plastic deformation in metals but is actually shear driven, or "quasi-plastic", without the presence of flow. Transgranular or intergranular shear fractures originate from high compression under the contact allow deformation and create the dimpled appearance at the surface. This phenomenon has been studied at length by Lawn[10] as well as Evans and Wilshaw[12].

Each material tested had a threshold force below which dimpling did not occur. An image of the 500-μm indents on SiC-B is given in Fig. 3(b). The dimples are around 40 μm in diameter or about 10 times the grain size. The actual maximum loads for each indent are also shown. It can be seen in this case that yielding consistently occurred for loads above 29 N, whereas it was elastic at lower forces. The additional tests on SiC-B and the other grades also showed the same level of consistency of apparent yield loads. The loads were nearly identical between different samples of the same material.

The SiC-HPN's Y_{App} was about 10% higher (~ 14 GPa vs. ~12.4 GPa) than that of the other three SiC grades, a statistically significant difference. The other three SiCs have similar yield strength values to one another. The measured average apparent yield loads and the calculated yield strengths determined from Eq. 4 are given in Table II. The low standard deviations show the high degree of consistency in these tests. SiC-HPN's higher Y_{App}, combined with its high resistance to ring crack initation[13] is consistent with the findings of Lundberg[14] in which SiC-HPN had the highest threshold velocity in reverse ballistic tests. Further work will examine the possible link between yield strength and the Hugoniot Elastic Limit (HEL)

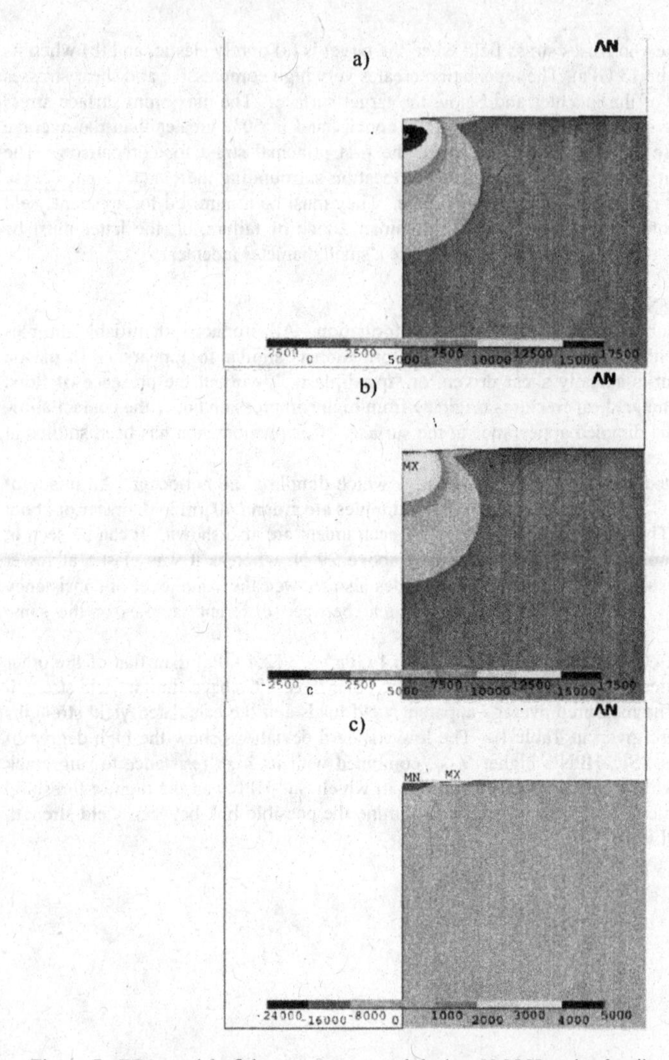

Figure 7. FEA model of the stresses created during 50 N Hertzian loading of SiC (E=450 GPa and ν=0.16) with a 500-µm-diamond indenter. (a) Purely elastic case, and (b) if yielding occurs with Y=13 GPa. The first principal stresses in the elastic case are shown in (c). Legend units are in MPa.

It is important to note that this study's spherical indentation is not merely a measure of material hardness, but in fact a parameter representing the apparent yielding behavior. The Knoop hardness of the four SiCs tested was shown in Fig. 4 and no significant difference exists between the average Knoop hardness of the four materials. However, the difference in Y_{App} is quite apparent as shown in Table II. This illustrates the discriminating value of this test method for comparing candidate materials because it is able to elucidate a statistically significant difference between contact damage resistances; a difference not exploited with Knoop hardness testing.

Table II. Measured average apparent yield strengths of the four SiC grades. Averages of 16 tests per grade and +/- values are 1 standard deviation.

Grade	Average Apparent Yield Load (N)	Apparent Tresca Yield Stress (GPa)
SiC-B	25.5 ± 0.90	12.3 ± 0.15
SiC-N	25.6 ± 0.85	12.5 ± 0.14
SiC-HPN	35.9 ± 0.96	14.0 ± 0.04
SiC-SC-1RN	26.6 ± 1.39	12.6 ± 0.22

SUMMARY

Apparent yield strengths of four hot-pressed silicon carbides (SiC-B, SiC-N, SiC-HPN, and SiC-SC-1RN) were estimated using diamond spherical or Hertzian indentation. SiC-HPN had an apparent yield strength of 14 GPa and that was approximately 10% higher than the others. This trend is consistent with observed threshold velocity ballistic behavior. The developed test method was robust, simple and quick to execute, enabling the acquisition of confident sampling statistics. A 500-µm-diameter indenter was effective at initiating dimpling before ring cracking initiated giving confidence that the initiation of an apparent yielding process was not complicated by pre-existing fracture. The compressive force necessary to initiate apparent yielding was identified postmortem using differential interference contrast (or Nomarski) imaging with an optical microscope. The discrimination in Y_{App} shows that this test method could be valuable because systematic Knoop hardness testing showed no differences among the four SiC grades.

ACKNOWLEDGEMENTS

The authors wish to express sincere appreciation to the: U.S. Army RDECOM-TACOM's D. Templeton for sponsoring this work, Swedish Defense Research Agency's P. Lundberg for providing the SiC materials, University of Tennessee's K. Johanns for the hardness measurements, ORNL's E. Kenik and H. -T. Lin for the SEM imaging, the US ARL's J. LaSalvia and J. Swab for their insights, and ORNL's J. G. Hemrick and C.-H. Hsueh for their review of the manuscript and their useful suggestions. Research sponsored by WFO sponsor US Army Tank-Automotive Research, Development and Engineering Center under contract DE-AC05-00OR22725 with UT-Battelle, LLC.

REFERENCES

[1] G. R. Johnson and T. Holmquist, "Response of Boron Carbide Subjected to Large Strains, High Strain Rates, and High Pressures" *J. App. Phys.* , **85,** 8060-73 (1999).

[2] R. Mouginot, "Blunt or Sharp Indenters: A Size Transition Analysis" *J. Am. Ceram. Soc.*, **71**, 658-61 (1988).

[3] K. L. Johnson, J. J. O'Connor, and A. C. Woodward, "The Effect of the Indenter Elasticity on the Hertzian Fracture of Brittle Materials," *Proc. R. Soc. London., Math. and Phys. Sci.*, **334**, 95-117 (1973).

[4] A. A. Wereszczak, K. E. Johanns, W. L. Daloz, and O. M. Jadaan, "Hertzian Ring Crack Initiation in Hot-Pressed Silicon Carbides," in preparation, 2008.

[5] D. Tabor, *Hardness of Metals*, Clarendon, Oxford, U.K., 28-49 (1951).

[6] A. A. Wereszczak and K. E. Johanns, "Spherical Indentation of SiC," *Cer. Engrg. Sci. Proc.*, **27**, 43-57 (2006).

[7] A. A. Wereszczak, "Elastic Property Determination of WC Spheres and Estimation of Compressive Loads and Impact Velocities That Initiate Their Yielding and Cracking," *Ceram. Engrg. Sci. Proc.*, **7**, 211-223 (2006).

[8] H. Li and R.C. Bradt, "The Microhardness Indentation Load/Size Effect in Rutile and Cassiterite Single Crystals," *J. Mater. Sci.*, **28**, 917-926 (1993).

[9] K. L. Johnson, *Contact Mechanics*, Cambridge University Press, London, U.K., 153-55 (1985).

[10] B. R. Lawn, "Indentation of Ceramics with Spheres: A Century after Hertz," *J. Am. Ceram. Soc.*, , 1977-94 (1998).

[11] R. M. Davies, "The Determination of Static and Dynamic Yield Stresses Using a Steel Ball" *Proc. R. Soc. London. Math. and Phys. Sci.*, **197**, 416-432 (1949).

[12] A.G Evans and T. R. Wilshaw, "Quasi-plastic Solid Particle Damage in Brittle Materials," *Acta Metall.*, **24**, 939-56 (1976).

[13] A. A. Wereszczak and R. H. Kraft, "Instrumented Hertzian Indentation of Armor Ceramics," *Cer. Engrg. Sci. Proc.*, **23**, 53-64 (2002).

[14] P. Lundberg and B. Lundberg, "Transition Between Interface Defeat and Penetration for Tungsten Projectiles and Four Silicon Carbide Materials," *Int. J. Impact Engrg.*, **31**, 781-92 (2005).

[15] R. F. Cook and G. M. Pharr, "Direct Observation and Analysis of Indentation Cracking in Glasses and Ceramics," *J. Am. Ceram. Soc.*, **73**, 787-817 (1990).

[16] W. C. Oliver and G. M. Pharr, "An Improved Technique for Determining Hardness and Elastic Modulus Using Load and Displacement Sensing Indentation Experiments," *J. Mater. Res.*, **7**, 1564-83 (1992).

MICROSTRUCTURAL EXAMINATION AND QUASI-STATIC PROPERTY DETERMINATION OF SINTERED ARMOR GRADE SiC

Memduh V. Demirbas, Richard A. Haber, Raymond E. Brennan
Rutgers University
Ceramic and Materials Engineering
607 Taylor Road
Piscataway, NJ 08854

ABSTRACT
Uniform microstructures with a low percentage of well distributed porosity could possibly demonstrate high ballistic strength; therefore, it is of interest to estimate the parameters that define the spatial arrangement of defects. This aspect of microstructures was investigated in a sintered armor SiC tile. A set of several tiles had been previously scanned by ultrasound and one tile from that set was selected for further evaluation. High and low amplitude regions in the C-scan image were examined using diced pieces from those areas. Acoustically varied regions were studied by optical microscopy. Segmentation of these artifacts and image analysis were performed. Initially, results on size of features were obtained. Subsequent spatial data analysis was performed to quantify microstructures in terms of homogeneity. Nearest neighbor distance distributions were utilized to study spatial distribution of artifacts. Quasi-static property measurements were employed to confirm microstructural findings. Correlations with ultrasound results were obtained and will be presented.

INTRODUCTION

The reaction of a ceramic material in the case of a ballistic event is complex and several parameters come into play in the ballistic performance of a material[1]. Defects are one of those parameters since they are the root causes of mechanical failure[2]. The emphasis will be given to "location" of defects in this paper as spatial distribution could be particularly important. The methodology that will be explained here is an approach to the particular problem of distinguishing samples with homogenous defect distributions from the ones where grouping, clustering or inhomogeneities in defect locations occur.
The perception in the armor community is that hardness is one of the key parameters for prediction of ballistic performance. It will be used as a gauge in order to verify microstructural examinations[3-6].
The opportunity for rapid evaluation in ultrasound is very attractive for armor tile inspection. Therefore, it is important to validate ultrasound results using destructive techniques. This paper focuses on understanding of contrasting ultrasound signals using microscopy and hardness tests. High frequency ultrasound was employed as this type of transducers provides the opportunity to detect smaller features[7].

EXPERIMENTAL PROCEDURE

The areas of interest were examined at a low magnification of ×100. Image analysis was applied to obtain quantitative information on the microstructures. Image Processing Toolkit 5.0 (Reindeer Graphics, Inc., Asheville, NC) was used for image analysis. The details of this procedure were explained in earlier papers by Demirbas and Haber[8,9].
For the ultrasound part of the study, the tile was scanned using a 75 MHz transducer. This facilitates higher resolution in order to detect smaller artifacts. The tile is immersed into a tank filled with water in order to obtain a small acoustic impedance mismatch to generate a strong signal. The transducer, which is connected to a stage that could move in x, y, and z directions, scans the entire surface of the tile, collecting signals which are assigned to x and y coordinates. The result is presented as C-scan ultrasound images.

RESULTS

Microstructural Assessment

A C-scan image of the sintered SiC tile is given in Fig. 1. The color scale represents the amplitude of ultrasound signals in milivolts (mV). The blue areas are low amplitude or acoustically poor regions that are segregated on two sides of the tile. The tile was cut and diced from those regions as shown between dashed lines in Fig. 1. The pieces from these two cuts represent "bad" regions. Another cut was made from the middle part of the tile which had predominantly shades of yellow and red regions. The pieces from this cut represent "good" regions.

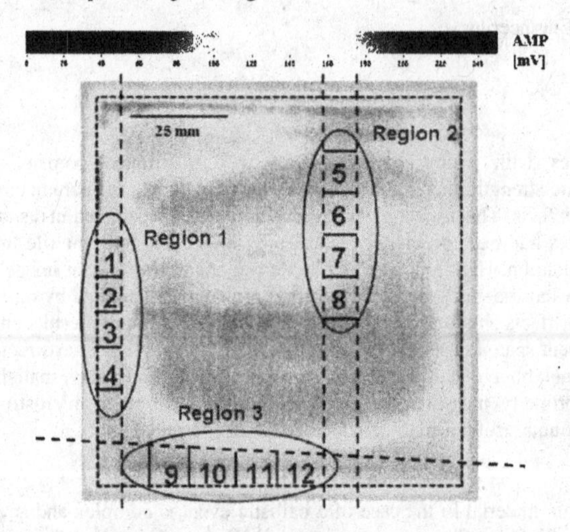

Figure 1. C-scan image of ultrasound tile

There were 12 pieces in total. Numbers 1 to 4 and 9 to 12 were from "bad" regions and numbers 5 to 8 were from "good" regions. Six of these of twelve pieces, namely N1, N4, N5, N8, N9, and N12, were studied using optical microscopy. N1 and N4 make up Region 1, N5 and N8 make up Region 2 while Region 3 consists of N9 and N12.

Optical microscopy was employed since low magnification of ×100 was sufficient and there was no need to use scanning electron microscopy. The purpose of using lower magnification was to cover as much area as possible in order to create a more valid comparison with ultrasound. An example image is given in Fig. 2. Ten pictures from each sample were combined to obtain the image in Fig. 2. Although it is a little tough to see the features clearly, they are presented this way in order to give an idea of the total area examined. The area covered is approximately 0.5 cm^2, which corresponds to almost half of the surface of the sample. This is a significantly large area that could provide insight to the interpretation of the acoustically poor regions.

Figure 2. Optical image of N1 at ×100

Quantitative information was obtained after the image analysis procedure was applied. The number of artifacts, density calculated by image analysis, average defect size, standard deviation of defect size and the maximum defect size are all given in Table I. Average defect size values are very close to each other as they vary between 4.46 μm and 4.63 μm. The one with the smallest average defect size is N12, which is from the "bad" region. However, in the descending order from the one with the smallest defect size to the largest one, the next two samples are N5 and N8, which are from the "good" region.

Table I. Results from image analysis on optical images

	1	4	5	8	9	12
No. of artifacts	14754	14455	15452	16445	14615	15359
Density (gr/cm^3)	3.185	3.187	3.187	3.187	3.188	3.188
Ave. Def. Size (μm)	4.63 ± 2.20	4.58 ± 1.93	4.52 ±1.71	4.47 ±1.65	4.63 ± 1.79	4.46 ±1.62
Max. Def. Size (μm)	63.10	35.58	25.54	21.95	27.76	38.41

Defect size distributions from both "good" and "bad" regions are given in Fig. 3. Curve fitting was performed using a version of the power law. The goodness of fit, R^2, values are all reasonably high with values above 0.91. Defect size distributions commonly seem to follow the power law[10-13]; therefore the current results are noteworthy. Another remark is the largest sized features found in samples from the both regions. The curve ends at below 30 μm for N5 from the "good" region while the curve for N1 goes beyond 60 μm.

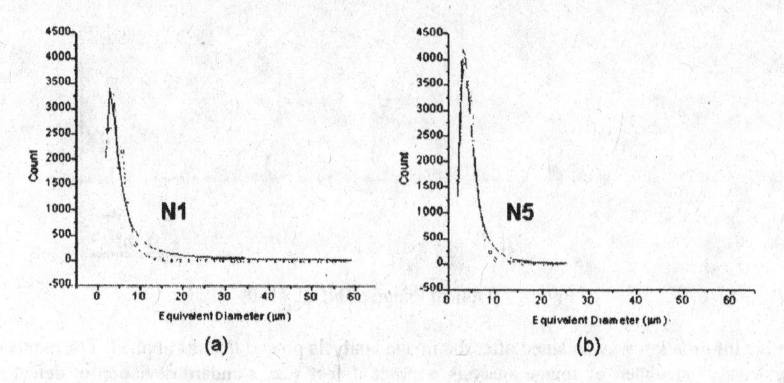

Figure 3. Defect size distributions from (a) N1 and (b) N5

After obtaining information related to size on artifacts, spatial distribution of artifacts were examined. Nearest neighbor distance distributions enables us to determine the uniformity of a microstructure by comparing the distributions to that of a complete random case. Q-V maps are utilized, where;

$$Q = \frac{\mu_o}{\mu_e} = \frac{\text{observed mean nearest - neighbor distances}}{\text{expected mean nearest - neighbor distances}}$$

$$V = \frac{\text{var}_o}{\text{var}_e} = \frac{\text{observed variance of nearest - neighbor distances}}{\text{expected variance of nearest - neighbor distances}}$$

the expected mean and variance are the expected mean and variance for a random distribution. Different types of spatial distributions can be categorized by the following guidelines: (a) random distributions, $Q \approx 1$ and $V \approx 1$; (b) regular distributions, $Q > 1$ and $V < 1$; (c) clustered distribution, $Q < 1$ and $V < 1$; and (d) random distribution with clusters, $Q < 1$ and $V > 1$[14].

Q-V map, given in Figure 4, shows some remarkable results as the average of data points for three out of six samples fall into the region of "clustering in a random background". The samples from the "good" region and N4 fall in the confidence interval of a "random" distribution. These results could be very significant in terms of explaining the differences observed in an ultrasound C-scan map. Clustering of artifacts could be an important factor among other possible reasons for variations.

So far, we have looked for deviations from "random" distribution, which could present itself as some degree of clustering of artifacts. Since, a certain degree of clustering was observed, it was important to actually locate the clusters. This will help us see the clusters in acoustically poor regions.

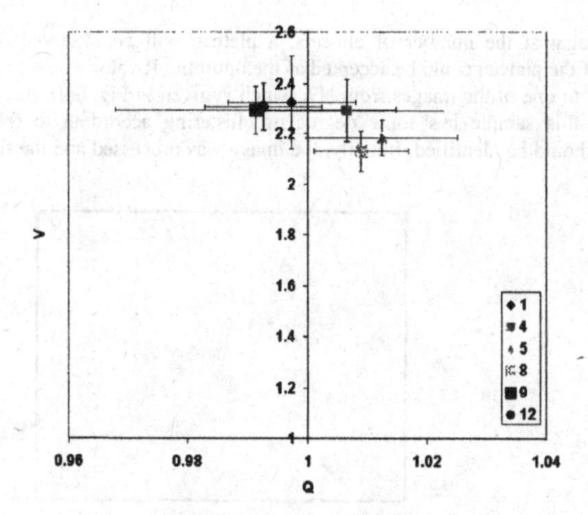

Figure 4. Q-V plot for all the samples

A method that was initially used by Anson and Gruzleski[15] was implemented into this study. In order to find clusters on images, the concept of *limiting interevent distance (R)* will be introduced. This is the radius of a cluster which contains all the features, or events, inside a circle. All the features outside this circle of radius R, are not a part of the cluster. The interevent distance is shown in Fig. 5.

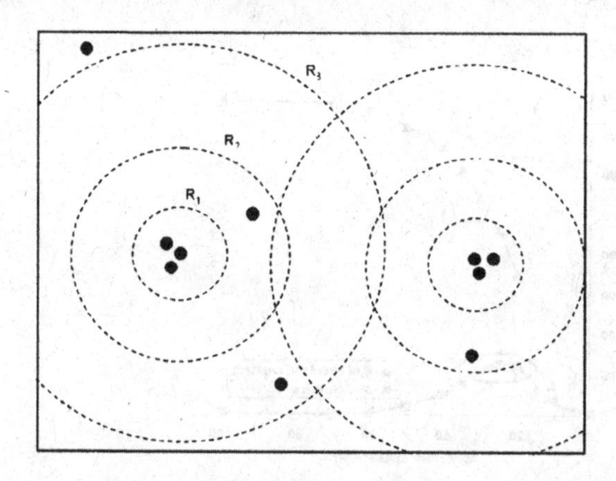

Figure 5. Illustration of limiting interevent distance (R)

The value of R must be determined carefully in order to obtain a reasonable representation of clustered and non-clustered regions. Anson and Gruzlezski[15] suggest a method, where, different values are assigned to R, from zero to very large values, until all the features in the image will be a part of one

cluster. If R is plotted against the number of clusters, a plateau will be observed. R value that corresponds to the start of the plateau could be accepted as the optimum R value.
This method was applied to one of the images from N9, which is given in Fig. 6(a). According to the previous measurements, this sample has some degree of clustering according to the Q-V map; therefore, these clusters should be identified. Initially, the image was processed and the final form can be seen in Fig. 6 (b).

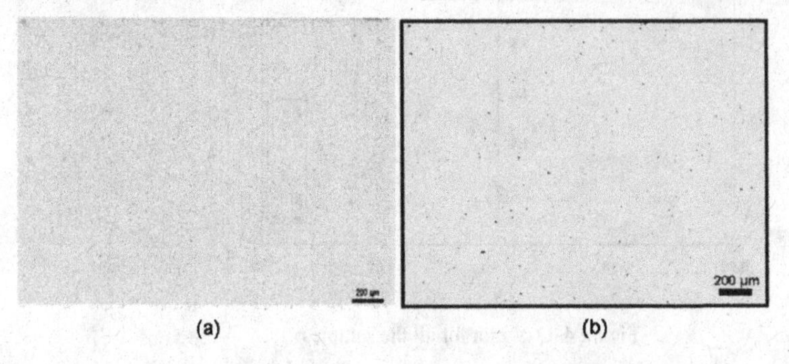

Figure 6. An optical image from sample N9

Using Image Processing Toolkit 5.0, R value was varied from zero to 100. In Fig. 7, the blue curve shows the number of clusters as R changes while the green curve demonstrates the number of features in clusters.

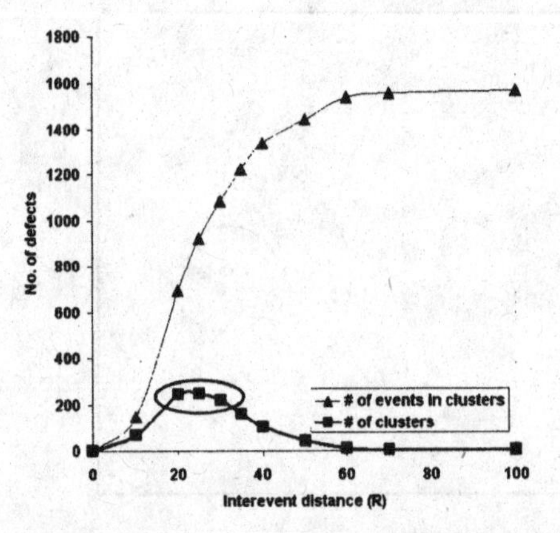

Figure 7. Determination of limiting distance interevent distance *(R)*

The plateau that was mentioned above can be observed between the values of 20 µm to 30 µm. Therefore, 20 µm can be accepted as the limiting interevent distance, R, according to the spatial distribution of features in this image. Fig. 8(a) shows the clusters in red, omitting the other features in the binary image that have nearest neighbor distance values above 20 µm. All artifacts can be seen in Fig. 8(b). Clustered artifacts are shown in red once again. Finally, clusters are demonstrated in the original image in Fig. 8(c). This part of the study is important in the sense that clusters can be identified in the images. This will provide more insight in the subsequent experiments such as dynamic tests since knowing the actual location of clusters helps to determine the effect they have on the overall performance.

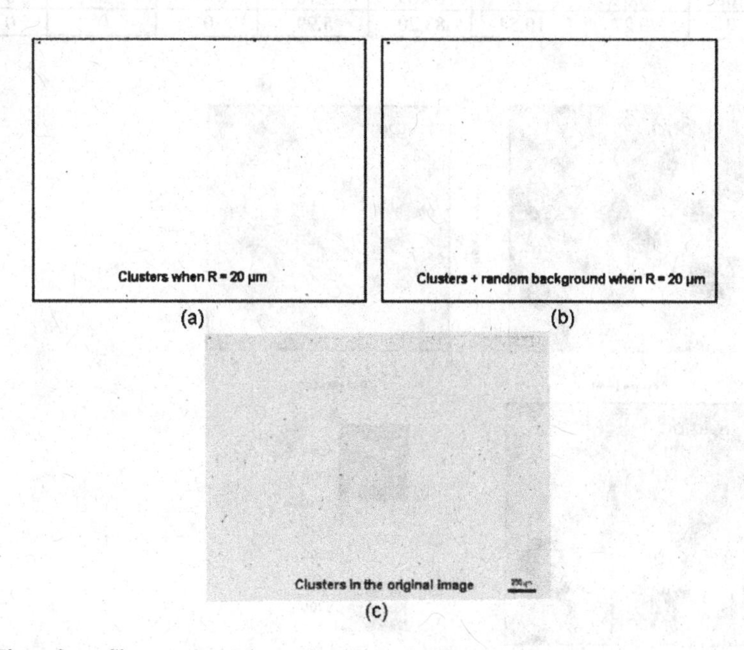

Figure 8. (a) Clusters when R is equal to 20 µm (b) Clusters superimposed on random background when R is equal to 20 µm (c) Clusters in the original image

Hardness Results

Hardness tests were conducted in order to test any correlation among microstructural findings, ultrasound results and a quasi-static property. Hardness contour maps were obtained by indentation of samples 100 times, forming a square of 10×10 indents. A Knoop indenter was used to indent the samples. The distance between each indent was kept at 0.5 mm. N2 represents Region 1, N6 represents Region 2 and N10 represents Region 3.

Contour maps are given Fig. 9 and the percentage of hardness values in each hardness range is given in Table II for each sample. The contour maps for N6 and N10 show more homogeneity while N2 seems to possess significant amount of hardness values from most of the ranges given in the scale. As it is shown in Table II, Region 2 has the highest percentage of hardness data with 84.69% in the most dominant data range, 1800-1900 KHN, represented by light green color. Region 3 is close with 83.20% but for the higher range, 1900-2000, it is much inferior to Region 2 with 5.99 % to Region 2's 9.16%. Region 1 has a wide spread which is quantified in Table II.

Table II. Percentage of Knoop hardness values in each hardness range

%	<1600	1600-1700	1700-1800	1800-1900	1900-2000	2000-2100	2100-2200	2200<
Region 1	0	0.23	20.77	68.35	9.67	0.98	0	0
Region 2	0	0.02	5.95	84.69	9.16	0.18	0	0
Region 3	0	0.27	10.54	83.20	5.99	0	0	0

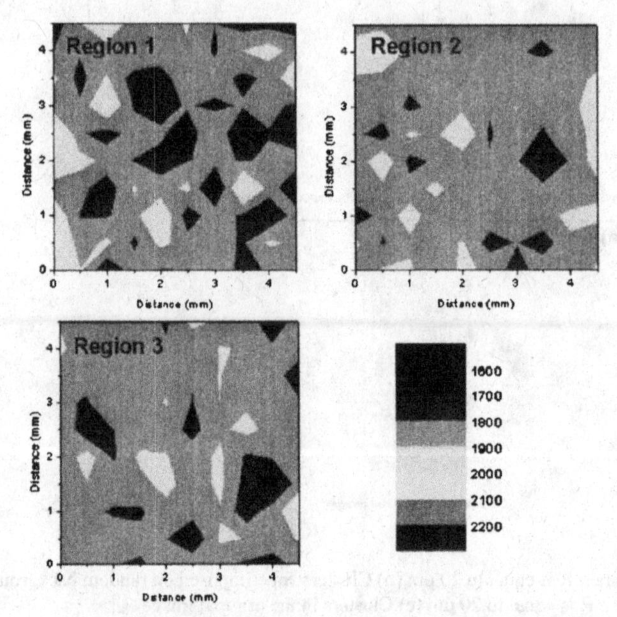

Figure 9. Hardness contour maps on samples from each region

SUMMARY

Three different sections of an armor grade sintered SiC tile were examined by using microstructural examination and quasi-static tests. The tile was cut and diced from acoustically poor regions and high amplitude regions. Nearest neighbor distance distributions were employed for spatial data analysis using optical micrographs. In this method, the distance between a pore and its closest neighbor is calculated and this is performed for all pores or other features in the field of view. In addition, a method for locating clusters on the images was developed. Quasi-static property measurements were employed to confirm microstructural findings. Contour maps were generated to show the change in hardness in terms of location.

Slight differences were observed between the samples from "good" and "bad" regions in terms of both microstructural and hardness results. Presence of larger size artifacts, a certain degree of clustering and a slightly lower average hardness values were observed in the "bad" samples. These could be an indication that both destructive and nondestructive testing point to similar tendencies and also, the results from ultrasound could be validated by using microscopy. There could be other factors that help us explain ultrasound results, but current findings are encouraging. Examination of several layers instead of one might reveal more on this subject and that type of study is currently in progress.

REFERENCES

[1]H. Luo, W. W. Chen, A. M. Rajendran, "Dynamic compressive response of damaged and interlocked SiC-N ceramics", Journal of the American Ceramic Society 89
[2]Introduction to Ceramics, 2nd Edition, by W.D. Kingery, H.K. Bowen and D.R. Uhlmann, John Wiley and Sons, New York, 1976
[3]D. Viechnicki, W. Blumenthal, M. Slavin, C. Tracy and H. Skeele, "Armor Ceramics – 1987", The Third Tacom Armor Coordinating Conference Proceedings
[4]E. Medvedovski, "Alumina ceramics for ballistic protection – Part 2", American Ceramic Society Bulletin, Vol. 81, No. 4 (2002) pp. 45-50
[5]M. Flinders, D. Ray, A. Anderson, and R. A. Cutler, "High-Toughness Silicon Carbide as Armor", J. Am. Cer. Soc. 88 [8] 2217–2226 (2005)
[6]J. Sternberg, "Material Properties Determining the Resistance of Ceramics to High Velocity Penetration", J. Appl. Phys., Vol 65(9), 1 May 1989
[7]R.E. Brennan, R. Haber, D. Niesz, J. McCauley and M. Bhardwaj, "Non-Destructive Evaluation (NDE) of Ceramic Armor: Fundamentals", Advances in Ceramic Armor, 26 (7) pp. 223-230 (2005)
[8]M.V. Demirbas and R. A. Haber, "Relationship of Microstructure and Hardness for Al_2O_3 Armor Material", Advances in Ceramic Armor II, CESP, p. 167-179, 2006
[9]M. V. Demirbas and R. A. Haber, "Defining Microstructural Tolerance Limits of Defects for SiC Armor", Ceramic Armor and Armor Systems II, Ceramic Transactions, 178, p. 109-122, 2005
[10]A. Jayatilaka, K. Trustrum, Statistical approach to brittle fracture. J Mater Sci 12:1426–30 (1977)
[11]Y. Zhang, M. Inoue, N. Uchida, K. Uematsu, "Characterization of processing pores and their relevance to the strength of alumina ceramics", Journal of Materials Research, Vol. 14, No. 8, 1999
[12]K. Uematsu, H. Abe, M. Naito, T. Hotta, N. Shinohara, "Flaw Size Distribution in High-Quality Alumina", Journal of the American Ceramic Society, Vol. 86, No. 6, pp. 1019-1021, 2003
[13]M.G. Gee, "Brittle fracture of hardmetals: Dependence of strength on defect size distribution", Int. J. Mech. Sci. 26, 85 (1984)
[14]P. A. Karnezis, G. Durrant and B. Cantor, "Characterization of Reinforcement Distribution in Cast Al-Alloy/SiC_p Composites" Materials Characterization 40, 97-109 (1998)
[15]J. P. Anson and J. E. Gruzlezski, "The Quantitative Discrimination between Shrinkage and Gas Microporosity in Cast Aluminum Alloys Using Spatial Data Analysis" Materials Characterization 43, 319-335 (1999)

QUANTITATIVE CHARACTERIZATION OF LOCALIZED AMPLITUDE VARIATIONS IN SILICON CARBIDE CERAMICS USING ULTRASOUND C-SCAN IMAGING

Raymond Brennan, James McCauley
US Army Research Laboratory
Aberdeen Proving Ground, MD 21005-5066

Richard Haber, Dale Niesz
Rutgers University
607 Taylor Road
Piscataway, NJ 08854-8065

ABSTRACT

Ultrasound C-scan imaging is a technique that is utilized to detect and locate inhomogeneities in various materials. One common application is the detection of localized amplitude variations which reflect apparent micro/macro structural inhomogeneties in ceramic materials. A method is being developed for extracting the most distinct variations from bottom surface reflected signal amplitude C-scan images and quantitatively evaluating them in terms of number, size, shape, and proximity. The data are used to determine size distributions for comparison of material integrity. This method is effective in quantitatively contrasting the number and distribution of localized amplitude variations detected in hot pressed silicon carbide (SiC) as compared to sintered SiC. While the number and size of significant individual variations detected in sintered SiC is much greater, the size distributions can be fit to same power law function with R^2 values of higher than 96% for each data set. Ultrasound detection and size distribution data can prove useful for ceramic material integrity comparison.

QUANTITATIVE THRESHOLD PROCEDURE

Ultrasound testing is a nondestructive method for transmitting acoustic signals through the bulk of a material to determine the presence of distinguishing variations. The resulting transmitted and reflected acoustic waves are collected in terms of either the time-of-flight, which is the travel time of the waves through the material, or the reflected signal amplitude, which is the intensity of the resulting signals. These data are spatially mapped to form a visual plot of acoustic differences throughout the bulk of the sample, which is referred to as a C-scan image. Ultrasound C-scan images can be used to nondestructively characterize localized amplitude variations that identify apparent micro/macro structural inhomogeneities[1-4]. These inhomogeneities may be the result of individual defects such as pores and inclusions or density variations in the materials.

A technique was developed for separating the most distinct individual amplitude variations from an ultrasound C-scan image and measuring key properties that could be categorized for determining size distributions. This was important for figuring out the number of distinguishable variations in a ceramic test specimen.

First, a C-scan image was generated by using a conventional ultrasound system to collect the raw data. The amplitude of the bottom surface reflected signal was assigned to a color scale based on the variation in signal intensity. This image was converted to grayscale and the brightness and contrast were optimized to generate the starting image (Figures 1 and 6). Next, a histogram curve of amplitude occurrences over the full grayscale range was used to choose the appropriate threshold value that could separate the most distinguishable localized variations. Thresholds were chosen from the left side of the curve to highlight the low amplitude regions and from the right side of the curve to highlight the high amplitude regions. Low reflected signal amplitude variations exhibited significant acoustic impedance (Z) mismatch compared to the bulk of the test specimen. Some examples of materials with lower acoustic impedance values than sintered SiC ($Z\sim37.5\times10^5$ g/cm^{2}s) include pores ($Z\sim0.0\times10^5$ g/cm^{2}s) and carbon (Z~1.5×10^5 g/cm^{2}s) inclusions. On the other hand, high reflected signal amplitude variations were a more rare case that included significant sized variations of higher density compared to the bulk of the test specimen. Since these exhibited a higher density and therefore a higher degree of ultrasound transmission, the bottom surface reflected signal amplitude values were higher than the sample average. Some examples of materials with higher acoustic impedance values than sintered SiC (Z~37.5×10^5 g/cm^{2}s) included Al_2O_3 (Z~43.0×10^5 g/cm^{2}s), iron (Z~45.4×10^5 g/cm^{2}s), and TiB_2 (Z~51.3×10^5 g/cm^{2}s) inclusions[1,4-7]. Low and high reflected signal amplitude variations were subject to the chosen thresholds and mapped separately (Figure 8). Erosion and dilation image processing steps were used on to remove noise and enhance the images for further evaluation[8].

The resulting map of high reflected signal amplitude variations was overlaid on the map of low reflected signal amplitude variations. This combined map showed all of the distinct amplitude variations that could be detected (Figure 9). The next step was to number and label them individually to determine their properties (Figures 3 and 10). After calibrating the scale of the image, the properties of each localized variation including area, perimeter, equivalent diameter, length, width, aspect ratio, symmetry, roundness, nearest neighbor distance, and minimum separation distance were calculated. Out of these parameters, the equivalent diameter was crucial for determining size. These values were collected and plotted in histogram form based on the number of occurrences of each size (y-axis) and the size range (x-axis). Different curve fit equations were plotted to find the highest correlation to the data and determine a representative equation that could describe the size distribution (Figures 5 and 12). Various maps were constructed to display localized amplitude variations within the chosen size ranges and provide a count of the number of distinct variations within each range (Figures 4 and 11).

INDIVIDUAL LOCALIZED AMPLITUDE VARIATIONS IN SILICON CARBIDE

This technique was applied to two different ceramic test specimens to determine the number of distinct variations and their properties including size distributions. Sample A was a hot pressed high density SiC specimen while sample B was a sintered SiC specimen containing 15 volume percent of a TiB_2 additive. While both samples had a length and width of 101.6 by 101.6 mm, sample A had a thickness of 19.1 mm while the thickness of sample B was 7.7 mm. While both samples had measured densities of 3.22 g/cm^3, sample A was a one-phase hot pressed sample close to theoretical density while sample B acted as a two-phase sintered system. Despite the additional porosity in the sintered sample, the high density second phase increased the overall measured density of the sample. Each of the C-scan images was collected at a step size of 0.05 mm, so the high resolution images included data sets of approximately 3.6 million data points.

For evaluation of sample A, a 125 MHz bottom surface reflected signal amplitude C-scan image was obtained. The color image was converted to grayscale and the brightness and contrast were optimized to bring out any distinct amplitude variations as shown in Figure 1. Since there were no large regions of low or high reflected signal amplitude, the sectioning step was unnecessary. The full range of grayscale values on the histogram curve went from approximately 50 to 250. Three different thresholds denoted low, medium, and high were selected for the low amplitude variations as well as the high amplitude variations, in which a low threshold collected a small number of variations and a high threshold collected a large number of variations. For the low amplitude variations, thresholds of 120, 130, and 140 were chosen as low, medium, and high while thresholds of 231, 201, and 171 were chosen as low, medium, and high for high amplitude variations at the other end of the spectrum. The medium thresholds for each amplitude range were chosen to represent each sample, since the low thresholds showed too few individual amplitude variations and the high thresholds showed too many. Maps of both the low amplitude and high amplitude variations were generated and combined to form a single image map as shown in Figure 2. Erosion and dilation were then performed on the combined image map. This screening method ensured that only the most prominent localized amplitude variations were being evaluated. Next, they were numbered individually as shown in Figure 3 and the properties were measured for the low amplitude, high amplitude, and combined maps.

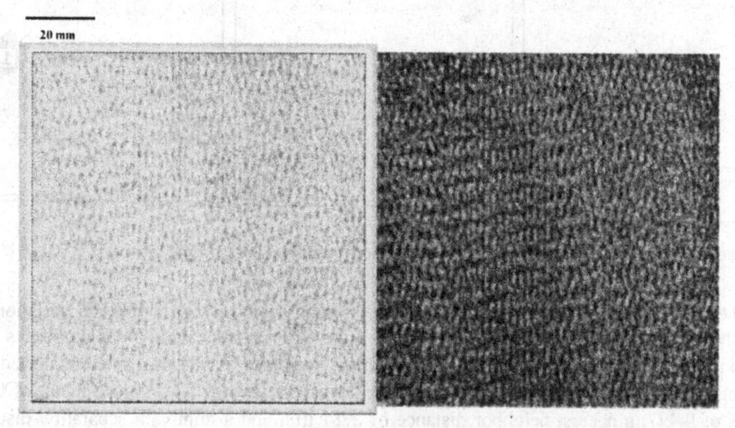

Figure 1. Bottom surface reflected signal amplitude color and gray scale images from sample A.

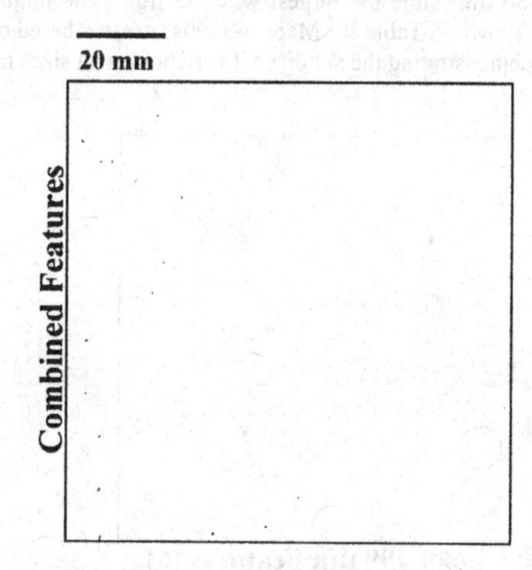

Figure 2. Threshold image of combined high and low amplitude variations from sample A.

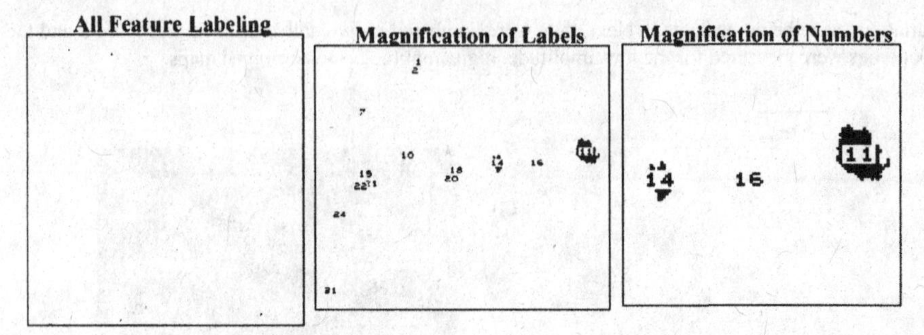

Figure 3. Labeling from combined map of sample A.

The data revealed that 131 low amplitude variations and 16 high amplitude variations were distinguished for a total of 147 in the combined map. The average individual properties for the combined image map included an area of 29,510 μm^2, a perimeter of 684 μm. an equivalent diameter of 159 μm, a length of 246 μm. a width of 156 μm. an aspect ratio of 1.457, a symmetry of 0.937, a roundness of 0.497. a nearest neighbor distance of 2781 μm, and a minimum separation distance of 2553 μm. The average parameters from each map are shown in Table I. The smallest localized amplitude variation size was 56 μm while the largest was 538 μm. The minimum and maximum parameters from each map are shown in Table II. Maps were also created based on size range, and an example is shown in Figure 4 demonstrating the six distinct variations with sizes ranging between 400-499 μm.

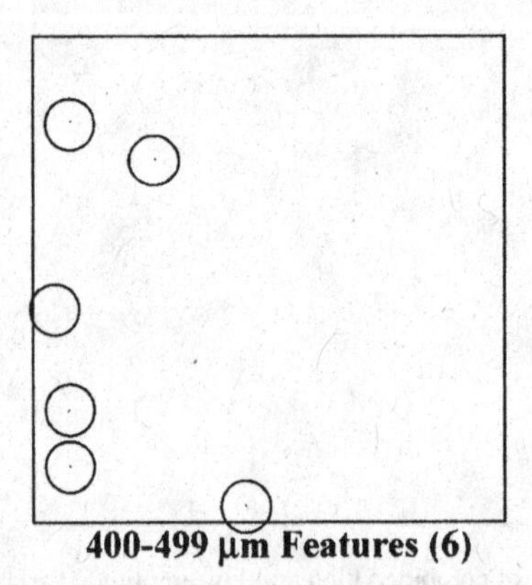

Figure 4. Example of size range map with six variations between 400-499 μm (circled) for sample A.

The equivalent diameter values were used to construct a size distribution curve. A scatter plot was first obtained based on the number of occurrences within specified size ranges. Next, curves were fit to the data. Since various forms of the power law function have been commonly used to describe size distributions[9-11], these functions were explored. The best curve fit was found to be a power law function, which had an R^2 correlation of 96.9% to the data set for sample A. The size distribution curve and fit are shown in Figure 5 in addition to the data describing the number of occurrences of within each range. The area-under-the-curve (AUTC) value for the size distribution curve was calculated as 8.363. The collected data represented amplitude variations found in the bulk of the material after nondestructive ultrasound testing.

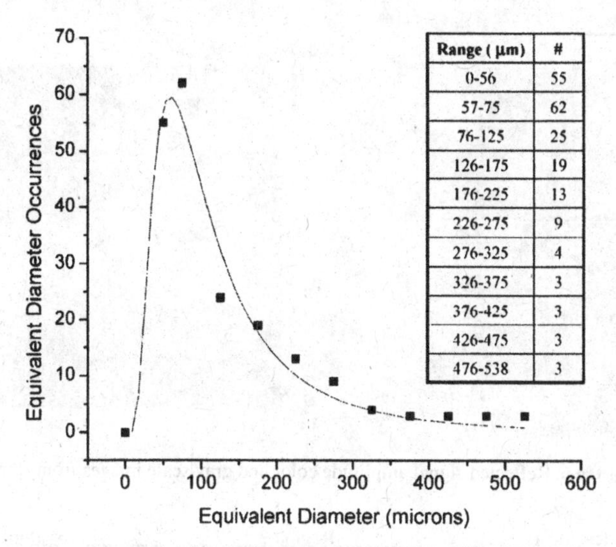

Range (μm)	#
0-56	55
57-75	62
76-125	25
126-175	19
176-225	13
226-275	9
276-325	4
326-375	3
376-425	3
426-475	3
476-538	3

Figure 5. Size distribution curve and fit for sample A.

For evaluation of sample B, the sintered SiC sample containing a TiB_2 additive, a 125 MHz bottom surface reflected signal amplitude C-scan image was also used. Again, grayscale conversion in addition to brightness and contrast adjustments were implemented, as shown in Figure 6. In this case, however, the sintered SiC material with the higher density additive revealed some low amplitude regions that had to be evaluated separately. The full range of grayscale values on the histogram curve went from approximately 50 to 250. When the initial threshold procedure was attempted, the low reflected signal amplitude region in the top right corner and some other smaller regions of lower amplitude were saturated so that no individual variations could be found. For this reason, the grayscale image was separated into three parts including the lower amplitude region from the top right corner (Region 1), the smaller regions with lower amplitude (Region 3), and the rest of the sample (Region 2), as shown in Figure 7. Thresholds for each of the regions were chosen in terms of both low and high reflected signal amplitude evaluation. For Region 1, the low amplitude thresholds were chosen as 120, 130, and 140 while the high amplitude thresholds were chosen as 155, 150, and 145. For Region 2, the low amplitude thresholds were chosen as 137, 147, and 157 while the high amplitude thresholds were chosen as 195, 185, and 170. For Region 3, the low amplitude thresholds were chosen as 134, 137, and 140 while the high amplitude thresholds were chosen as 160, 158, and 156. The medium thresholds were selected for each section since they provided sufficient data without drowning out too many amplitude variations or resulting in an excess of amplitude variations. After choosing the proper thresholds, the regions were merged to provide a single low amplitude image and a single high amplitude image as shown in Figure 8, and erosion and dilation were performed. Due to the high number of overall localized amplitude variations in sample B as compared to sample A, the defect size cutoff was limited to 99 μm as opposed to the previous minimum of 56 μm. Low amplitude and high amplitude maps were combined into a single map as shown in Figure 9. The combined map was much different than for sample A, as it contained a much greater number of overall variations, many of which were large in size.

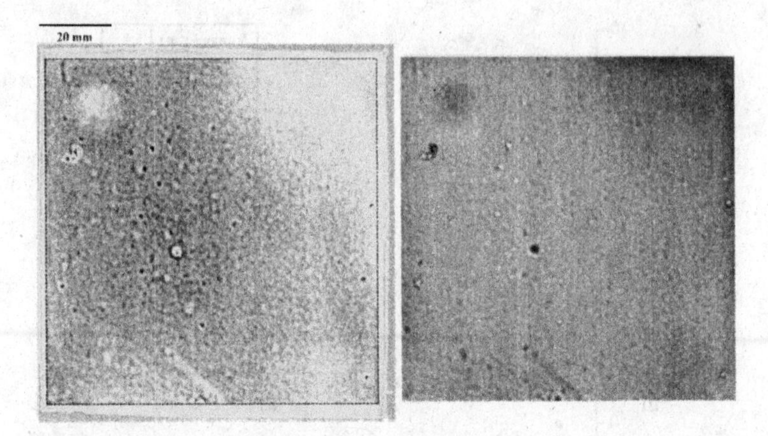

Figure 6. Reflected signal amplitude color and gray scale images from sample B.

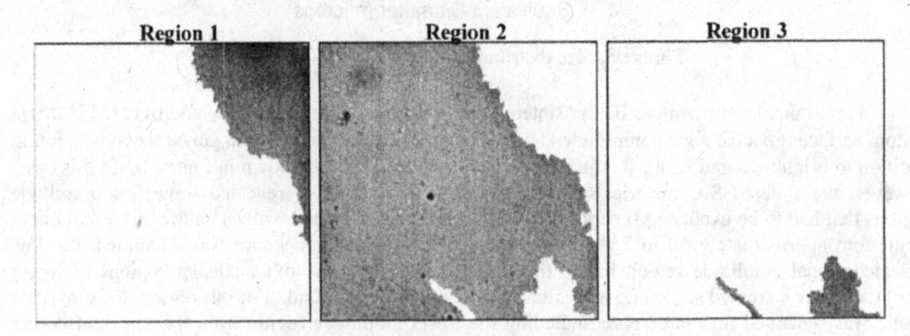

Figure 7. Regions chosen for threshold evaluation of sample B.

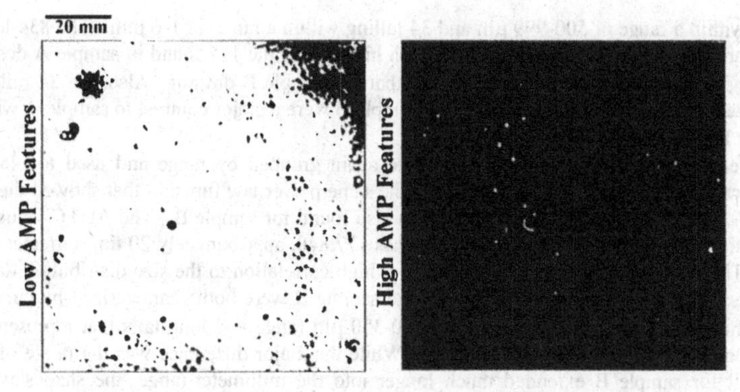

Figure 8. Threshold image selections of high and low amplitude variations from sample B.

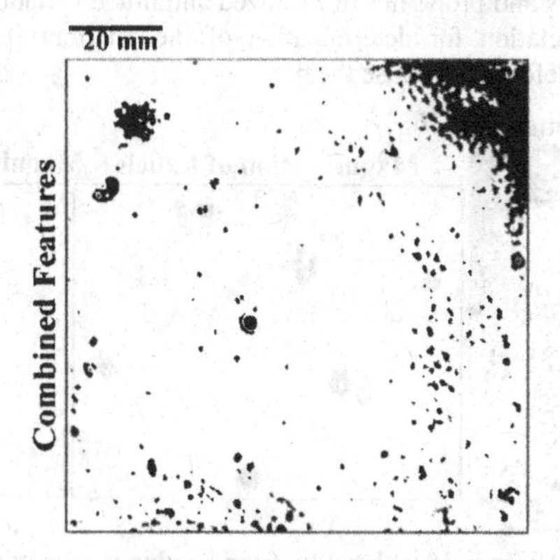

Figure 9. Threshold image of combined high and low amplitude variations from sample B.

The amplitude variations were numbered individually in Figure 10 and properties measured. The data sheet revealed that 769 low amplitude and 66 high amplitude variations were distinguished for a total of 835 in the combined map. The average individual properties for the combined image map included an average area of 1.96 x 10^5 μm^2, an average perimeter of 1615 μm, an average equivalent diameter of 332 μm, an average length of 472 μm, an average width of 316 μm, an average aspect ratio of 1.427, an average symmetry of 0.933, an average roundness of 0.566, an average nearest neighbor distance of 807 μm, and an average minimum separation distance of 411 μm. The full list of average values from low amplitude, high amplitude, and combined images is shown in Table III. The smallest low amplitude size detected was 99 μm while the largest was 5.5 mm. The full list of minimum and maximum values from low amplitude, high amplitude, and combined images is shown in Table IV. These millimeter-range localized amplitude variations greatly increased the average area, perimeter, equivalent diameter, length, and width values as compared to sample A. The large number of high amplitude variations was attributed to higher density TiB_2 additives that were dispersed throughout sample B. Maps were created based on size ranges, and two examples are shown in Figure 11, with

134 falling within a range of 500-999 μm and 34 falling within a range of 1-6 mm. The 834 localized amplitude variations found in sample B were much higher than the 147 found in sample A despite the fact that sample A included the 50-99 μm range that the sample B did not. Also, the 34 millimeter-range localized amplitude variations detected in sample B were a major contrast to sample A which did not have any larger than 538 μm.

The equivalent diameter values were once again grouped by range and used to plot a size distribution curve as shown in Figure 12. Using the same power law function that showed the best fit for sample A, a high R^2 correlation of 96.3% was also found for sample B. The AUTC value for the size distribution curve of sample B was calculated as 177,840, approximately 20 times greater than for sample A. The power law function demonstrated a high correlation to the size distribution data from both samples. The histograms from sample A and sample B were both characterized by curves with the highest number of occurrences within the 150-350 μm range and long tails that represented few occurrences of the largest amplitude variations. While the major difference was the range of values, since the tail for sample B extended much longer into the millimeter-range, the shapes were still comparable. This technique proved to be valuable for describing large volumes of ceramic test specimens in terms of the sizes and properties of localized amplitude variations. It could also prove to be valuable for further correlation for determination of the relationship between these types of amplitude variations and the defects that cause them.

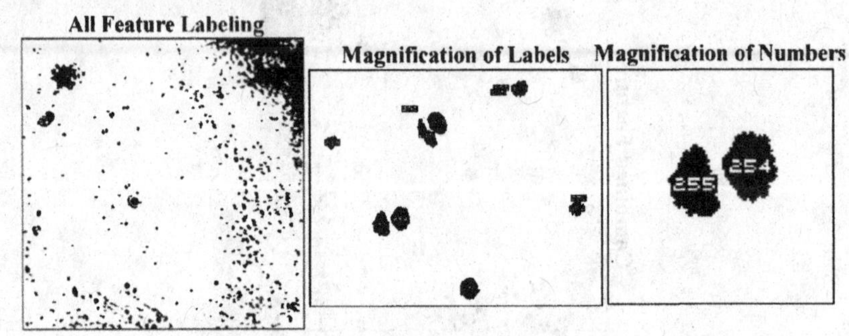

Figure 10. Individual labeling from combined map of sample B.

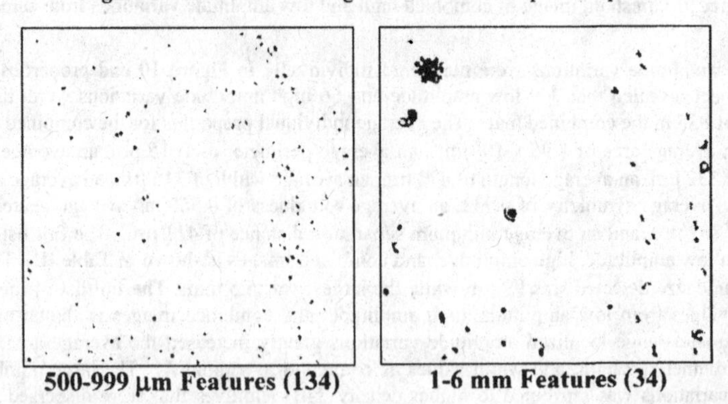

Figure 11. Size range maps for sample B including number of amplitude variations (in parentheses).

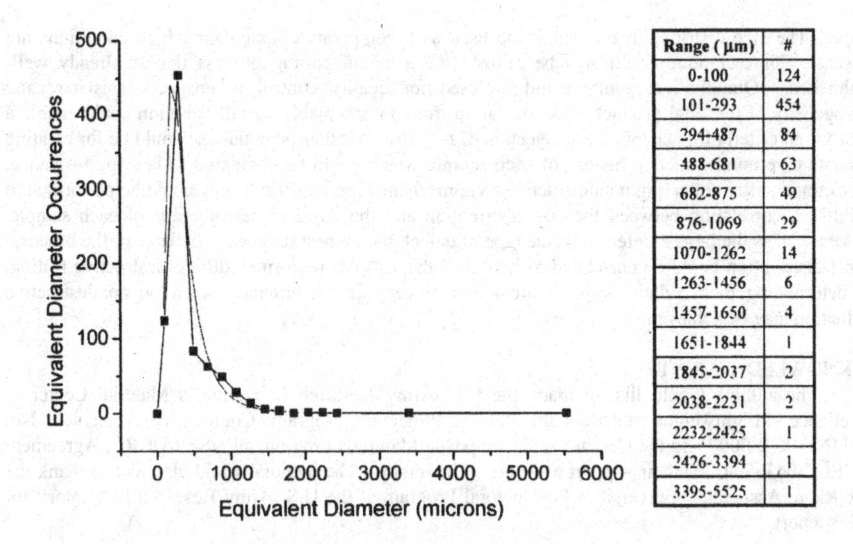

Range (μm)	#
0-100	124
101-293	454
294-487	84
488-681	63
682-875	49
876-1069	29
1070-1262	14
1263-1456	6
1457-1650	4
1651-1844	1
1845-2037	2
2038-2231	2
2232-2425	1
2426-3394	1
3395-5525	1

Figure 12. Size distribution curve and fit for sample B.

CONCLUSIONS

A technique was developed for effectively isolating the most distinct individual localized amplitude variations from an ultrasound C-scan image and characterizing them in terms of number, size, shape, and proximity. Size distribution data was also collected for comparison of the ceramic material integrity. This method was applied to high frequency ultrasound C-scan images of sample A, a hot pressed SiC sample, and sample B, a sintered SiC sample containing a high density additive. The quantitative threshold technique revealed that sample A showed 147 distinct individual variations compared to 834 for sample B. While sample A did not contain any localized amplitude variations larger than 538 μm, sample B contained 34 amplitude variations larger than 1 mm. Despite the large difference in number and size, the distribution data followed the same power law function for both SiC test specimens, with high R^2 correlation values of over 96% for each curve. By establishing a common function that described size distribution for both materials, the magnitude of the data, as described by the AUTC values became the most distinguishing factor. In this study, the sintered Sample B containing 15 volume percent of a high density additive exhibited an AUTC value that was 20 times larger than the hot pressed sample A with no additive.

The localized amplitude variations that were detected during ultrasound C-scan imaging indicated the presence of apparent micro/macro structural material inhomogeneities. The variation in amplitude could be attributed to individual defects such as pores or inclusions. It could also be attributed to local density variations. After establishing a direct correlation between the localized amplitude variations and specific individual defects, the aforementioned quantitative assessment technique could have many potential applications. When determining final material ingredients and conditions for a specific process, quantitative localized amplitude variation data could be compared for each type of additive or each percentage of second phase addition to provide the optimum material recipe. The size distribution data could be used as a comparative gauge for which conditions are chosen. This technique could also be utilized for a manufacturing process that is already well-established. Quantitative results could be used for quality control to ensure consistency and homogeneity of the final product. Any deviation from an acceptable size distribution could enable a go or no-go criteria for acceptance or rejection of the part. Another potential use could be for creating a quantitative nondestructive history of each sample which could be compared to later performance. For example, after collecting nondestructive data, mechanical or ballistic testing could be performed to establish a correlation between the size distribution and the physical performance of each sample. Factors such as the largest defect size, the type of defect, the highest number of defects, or the broadest defect distribution could be correlated to how well the samples performed during destructive testing. By determining these relationships, a prediction of ceramic performance based on nondestructive evaluation may be achieved.

ACKNOWLEDGEMENTS

The authors would like to thank the U.S. Army Research Laboratory's Material Center of Excellence - Lightweight Materials for Vehicle Protection Program, Cooperative Agreement No. W911NF-06-2-0007 and the Ceramic and Composite Materials Program an NSF I/UCRC, Agreement No. EEC-0436504, for their support at Rutgers University. The authors would also like to thank the Oak Ridge Associated Universities Postdoctoral Program at the U.S. Army Research Laboratory for their support.

REFERENCES

[1]P.E. Mix, *Introduction to Nondestructive Testing*, John Wiley & Sons, 104-153 (1987).

[2]J. Krautkramer, H. Krautkramer, *Ultrasonic Testing of Materials*, Springer-Verlag, (1990).

[3]D.E. Bray, R.K. Stanley, *Nondestructive Evaluation: A Tool in Design, Manufacturing, and Service*, CRD Press, Inc., 53-178 (1997).

[4]M.C. Bhardwaj; "Evolution, Practical Concepts and Examples of Ultrasonic NDC", *Ceramic Monographs – Handbook of Ceramics*, **41**, 1-7 (1992).

[5]National Institute of Standards and Technology Property Data Summaries for Advanced Materials (http://www.ceramics.nist.gov/srd/summary/advmatdb.htm/)

[6]S.J. Schneider, *Engineered Materials Handbook: Ceramics and Glasses, Vol. 4*, ASM International (1991).

[7]R.G. Munro, "Material Properties of Titanium Diboride", *Journal of Research of the National Institute of Standards and Technology*, **105**, 709-720 (2000).

[8]C.-H.H. Chu, E.J. Delp, "Impulsive Noise Suppression and Background Normalization of Electrocardiogram Signals using Morphological Operators", *IEEE Transactions on Biomedical Engineering*, **36** [2], 262-273 (1989).

[9]M.P. Bakas, "Analysis of Inclusion Distributions in Silicon Carbide Armor Ceramics", *Ph.D. Thesis*, Rutgers University (2006).

[10]A.M. Korsunsky, K. Kim, L.R. Botvina, "An Analysis of Defect Size Evolution", *International Journal of Fracture*, **128**, 139-145 (2004).

[11]H. Abe, M. Naito, T. Hotta, N. Shinohara, K. Uematsu, "Flaw Size Distribution in High-Quality Alumina", *Journal of the American Ceramic Society*, **86** [6], 1019-1021 (2003).

Table I. Average low, high, and combined threshold map values from sample A.

Average Feature Parameters

Image	Area [μm²]	Perimeter [μm]	Equivalent Diameter [μm]	Near. Neighb. Dist. [μm]	Min. Sep. Dist. [μm]
Low AMP	29,542	685	157	2676	2502
High AMP	29,375	669	182	9564	8082
Full	29,510	684	159	2781	2553

Additional Average Feature Parameters

Image	Length [μm]	Width [μm]	Aspect Ratio	Symmetry	Roundness
Low AMP	247	153	1.475	0.935	0.486
High AMP	231	186	1.217	0.958	0.630
Full	246	156	1.457	0.937	0.497

Table II. Minimum and maximum low, high, and combined threshold values from sample A.

Minimum and Maximum Selected Parameters

Image	Min. Area [μm²]	Max. Area [μm²]	Equ. Diam. Min. [μm]	Equ. Diam. Max. [μm]
Low AMP	2500	227,500	56	538
High AMP	12,500	80,000	126	319
Full	2500	227,500	56	538

Table III. Average low, high, and combined threshold map values from sample B.

Average Feature Parameters

Image	Area [μm²]	Perimeter [μm]	Equivalent Diameter [μm]	Near. Neighb. Dist. [μm]	Min. Sep. Dist. [μm]
Low AMP	1.97E5	1620	327	826	427
High AMP	1.90E5	1560	388	3319	2872
Full	1.96E5	1615	332	807	411

Additional Average Feature Parameters

Image	Length [μm]	Width [μm]	Aspect Ratio	Symmetry	Roundness
Low AMP	467	311	1.430	0.933	0.567
High AMP	523	371	1.397	0.934	0.595
Full	472	316	1.427	0.933	0.566

Table IV. Minimum and maximum low, high, and combined threshold values from sample B.

Minimum and Maximum Selected Parameters

Image	Min. Area [μm²]	Max.Area [μm²]	Equ. Diam. Min. [μm]	Equ. Diam. Max. [μm]
Low AMP	7801	2.40E7	100	5525
High AMP	7801	1.91E6	100	1557
Full	7801	2.40E7	100	5525

GRAIN BOUNDARY ENGINEERING OF SILICON CARBIDE BY MEANS OF COPRECIPITATION

Steven Mercurio, Mihaela Jitianu, Richard A. Haber
Rutgers University
607 Taylor Rd.
Piscataway, NJ 08854

Abstract:

Coprecipitation is a promising method for homogeneously distributing sintering aids and enabling improved microstructural control in silicon carbide (SiC). The more uniform coating and finer particle size achieved through coprecipitation avoids the agglomeration developed from ball milling of micron sized additives. Rare-earth elemental and oxide additions have been found useful to further promote liquid-phase sintering and enable modification of the grain boundary phases present in silicon carbide-based ceramics. Compositions of 95 wt% SiC - 5 wt% (xRE_2O_3- yAl_2O_3), where molar ratios x:y were set at 3:2 and 4:1 respectively, were processed by coprecipitating aluminum and several rare earth (RE = La, Gd, Y) hydroxides directly into slurries of dispersed silicon carbide. The samples were subsequently green processed and sintered before characterization. After coating, zeta potential measurements showed that the electrokinetic properties of SiC powder in aqueous medium have been modified. The structure of the coating layer was evaluated by Fourier transformed infrared spectroscopy (FTIR). The size and distribution of the coating particles was investigated by scanning electron microscopy (SEM). Preliminary TEM examination confirms the presence of grain boundary phase in sintered samples.

Introduction:

Silicon carbide armor ceramics continue to display a large variability in ballistic performance from tile to tile and batch to batch. Much of this variance is believed to be tied to subtle microstructural differences developed from poor processing. Large, exaggerated grains, inclusions, impurities, poorly distributed sintering aids, and other processing artifacts all result in defects that can lead to lower sintered densities and microstructural features that limit ballistic performance[1].

Eliminating the defects present in silicon carbide tiles would offer a moderate improvement in mechanical properties and performance, but it is unlikely such a change would serve to defeat the higher energy threats of the future. Silicon carbide, as it is currently processed, is unlikely to be able to match the strength, Hugoniot elastic limit, or hardness performance of boron carbide or composite materials in the future. SiC does however offer advantages in the ease of processing and the presence of intergranular films, twinning, and stacking faults. In order to establish silicon carbide as a viable material system for the next generation of armor ceramics, it is necessary to understand and control these grain boundary characteristics so as to enable new fracture mechanisms and grain boundary behaviors. By carefully engineering the grain boundaries, properties that are believed to be favorable to ballistic performance such as effective plasticity and toughness can be enhanced and controlled, leading to improved silicon carbide armor.

Much previous work has been done on modified sintering aids and processing for silicon nitride and silicon carbide. Becher et al. explored the effect of rare earth elements in sintered silicon nitride and observed marked differences in anisotropic grain growth with the segregation of different rare earth elements[2]. Zhou et al. added rare earth sintering aids like yttria, ytterbium, lanthanum, and niobium to silicon carbide and observed a distinct core-rim microstructure[3]. They also correlated important differences in mechanical properties with the varying cationic radius of the rare earths. Previous work in this research group by Ziccardi and Mercurio utilized a proprietary alkylene amine based surfactant (HX3 – Huntsman Chemical, Austin, Texas) which contains a high percentage of carbon and cross linked boron to introduce the sintering aids in a well mixed and controlled manor[4, 5]. This method improved over adding the boron and carbon as powders or as a resin, and lead to the development of silicon carbide with increased hardness and lower excess carbon. Adding boron and carbon alone however, necessitates high sintering or pressing temperatures and does not develop any appreciable intergranular film. Lower processing temperatures would lead to reduced costs and greater flexibility in processing, so additives other than just boron and carbide were desired. Research on coupling metal or rare earth elements to the Huntsman additives proved to be unsuccessful so previously aluminum compounds were added as bulk powders. While this method improved upon the previous ones, a finer, more uniform addition was still desired.

Coprecipitation of aluminum and rare earth elements would allow for a fine coating to be deposited on the silicon carbide while not requiring much modification of the current wet processing route. Previous work by Yang et al. has shown that coprecipitating aluminum and yttrium onto Si_3N_4 resulted in improved homogeneity of the sintering additives versus dry mixing methods[6]. Albano and Garrido found improvements in green density and sintered density of Si3N4 through coprecipitation as a means of coating[7]. Limited work on coprecipitation in SiC has shown improved sintering kinetics, greatly decreased grain size, and improved mechanical properties[8]. The coprecipitation method is versatile in that it can be carried out in the presence or absence of various surfactants or additives. This work investigates the coating of SiC with aluminum and rare-earth hydroxides by coprecipitation, without using a surfactant and also in the presence of the Huntsman HX3 additive (H.C.Spinks Clay Company, Paris, TN). The coprecipitated samples were examined to confirm the presence of the coating on SiC particles. For this purpose, electrokinetic measurements, along with Fourier transform infrared spectroscopy (FTIR) and scanning electron microscopy (SEM) techniques have been employed. The presence of a grain boundary phase in the sintered samples was investigated by transmission electron microscopy (TEM).

Experimental Method:

The starting powders used in this study were α-SiC (UK Abrasives), yttrium nitrate hexahydrate ($Y(NO_3)_3 \cdot 6H_2O$, Alfa Aesar, Ward Hill, MA), aluminum nitrate ($Al(NO_3)_3 \cdot 9H_2O$, Acros Organics, NJ), lanthanum nitrate hexahydrate ($La(NO_3)_3 \cdot 6H_2O$,

Alfa Aesar, Ward Hill, MA), and gadolinium nitrate hexahydrate ($Gd(NO_3)_3 \cdot 6H_2O$, Alfa Aesar, Ward Hill, MA).

The primary composition chosen for this study was 95 wt% SiC and 5 wt% of sintering aids, with a controlled molar Al:Re ratio of either 3:2 or 4:1. Secondary compositions were prepared using the HX3 surfactant. It was expected that the surfactant would improve the adhesion of the coprecipitates to the treated SiC surface. The secondary composition of 94 wt% SiC, 3 % HX3, 2 wt% Al, 1 wt % La was chosen because of the success of previous work with ball milling and hot pressing that composition. The amounts of the initial nitrate precursors were controlled to yield the desired final compositions.

Composition	Re	Al:Re
95 wt % SiC, 5 % additives	Y	3:2, 4:1
	Gd	3:2, 4:1
	La	3:2, 4:1
95 wt % SiC, 2.5 % HX3, 2.5 % Al	--	--
94 wt % SiC, 3 % HX3, 2 % Al, 1 % La	La	2:1

Table 1: Compositions of Sintered SiC Samples

Sample preparation:

The silicon carbide was dispersed in distilled deionized water and ultrasonicated. Two solutions were then prepared, one of the aluminum and rare earth nitrates in DDI water and the other of tetramethylammonium hydroxide pentahydrate (TMAH) ($C_4H_{12}NOH \cdot 5\ H_2O$, Fisher Scientific, Fair Lawn NJ) in DDI water. The aluminum and rare earth solution was slowly added dropwise to the SiC slurry. The TMAH solution was then slowly dropped into the slurry while the pH was carefully monitored until the addition was complete. The completed solution was then allowed to age for 1 hour at a pH of 10.

After settling, the solution was centrifuged and the supernatant was separated. Fresh DDI water was added, and then the slurry was mixed and centrifuged again. This washing procedure was then repeated six times for each composition. The powder was then dried at 100°C and subsequently ground. Small discs 0.75 inches in diameter were dry pressed and loaded into a custom made graphite die coated with boron nitride to prevent sticking.

The samples were pressureless sintered under argon at 1800°C and 1900°C respectively with a 1 hour hold at temperature. The sintering cycle was not designed for achieving full density, but instead allow for the samples to be easily crushed and processed for characterization.

Sample characterization:

The electrokinetic measurements were carried out on a Brookhaven Zeta PALS instrument using suspensions of 0.01g/mL SiC-based powders prepared in KCl 10^{-3}M. Fourier-transformed infrared spectroscopy (FT-IR) analysis was performed using a Galaxy 5000 Mattson spectrometer, for which samples were prepared in KBr pellets. Scanning electron microscopy images were acquired on a FE-SEM LEO Zeiss 982 microscope using an acceleration of 5 kV. The TEM measurements have been carried out on a Transmission electron microscope TEM Topcon 002B. Finely powdered samples have been dispersed in isopropanol and then deposited on Formvar copper carbon coated grids and dried overnight. The images have been taken using an acceleration voltage of 200 kV.

Results and Discussion:

FTIR results for SiC, Al/La 3:2, and coated SiC- Al/La 3:2 before and after sintering confirm the presence of a coating on the SiC powder and the formation of a glassy phase upon firing at 1900°C (Figure 1). The spectrum from the uncoated SiC powder shows both the longitudinal optical (LO) and transverse optical (TO) phonon bands at 929 cm^{-1} and 861 cm^{-1} as expected for a lattice constructed by polar chemical bonds[9], bands assigned to stretching vibrations of Si-O-C and Si-O-Si, respectively. The band at 1023 cm^{-1} can be attributed to the Si-O asymmetric stretching vibration since it is similar in shape to the band observed in amorphous silica[9], indicating the presence of a silicon oxide layer on the SiC powder. The band at 761 cm^{-1} corresponds to the Si-C stretching vibration mode[10].

The spectrum of the coprecipitated metal hydroxides confirms the coprecipitation formed hydroxides as expected due to the presence of the bending vibration band of HOH at 1670 cm^{-1}. A shoulder observed at 554 cm^{-1} (not shown) was assigned to AlO_6 octahedra. The band at 803 cm^{-1} along with that at 1030 cm^{-1}, can be assigned to La-oxycarbonate species according to some authors[11]. On the other hand, the band at 1030 cm^{-1} can also be assigned to N-based organic residual groups on the hydroxides surface[12]. The band at 1385 cm^{-1} is attributed to absorbed NO_3^- from the starting nitrates on the surface[12]. The washing procedure could be expanded to remedy the residual NO_3^- problem.

FTIR of the system with the coprecipitated hydroxides on the SiC slurry shows the Si-C stretching vibration mode band at 761 cm^{-1}. This band shows no change from the uncoated SiC as expected. The band at 855 cm^{-1} appears to be slightly shifted towards lower wavenumbers as compared to the TO band at 861 cm^{-1} from the uncoated SiC, but actually the band may be an overlap between that at 803 cm^{-1} present in the coating hydroxides, and the unshifted TO band.

Upon annealing at 1800°C for 1 hour, the position of stretching vibrations of Si-C-O and Si-O-Si bands[13] is shifted towards higher wavenumbers after annealing (886 cm^{-1} with a shoulder at 942 cm^{-1}). The band of Si-C stretching vibration is shifted at 780 cm^{-1} after annealing, as well, as reported also by Oliveira et el[10]. These changes may represent a change in the polytype distribution of the SiC during annealing. The band developing at 830 cm^{-1} can be attributed to condensed AlO_4 tetrahedra identified in most aluminates[14], indicative of a starting formation of an aluminosilicate glassy phase. The absorption bands due to the presence of La(III) in the glassy phase are usually present in the spectral range of 400 – 500 cm^{-1} and could not be identified in our spectra[11], because of the frequency range of our spectrometer, which is up to 500 cm^{-1}.

The spectrum recorded after annealing of the sample at 1900°C is different than that recorded for the sample annealed at 1800°C. An infrared spectrum of silica-rare earth-alumina glassy phase is expected to show two bands at ~ 1100 cm^{-1} and 943 cm^{-1}, respectively, bands related to the number of non-binding oxygen atoms and rare-earth presence[15]. The shoulder in our spectra at 933 cm^{-1} is actually an overlap between the LO band of SiC and the latter above-mentioned vibration. The shoulder at 670 cm^{-1} is due to the stretching vibration of isolated tetrahedra found in silica-rare earth-alumina glass structure[15]. The band due to the stretching vibration of Si-C becomes sharper and more intense after annealing at 1900°C, as it was previously found by Alexeev et al[9] and it also shifts towards lower wavenumbers. This may be a possible indication of changing of the polytype of SiC in the presence of a glassy phase.

As observed in Figure 1, in all of the spectra, the strong LO and TO absorption bands of the SiC masked the other absorption bands, making full identification through FTIR difficult.

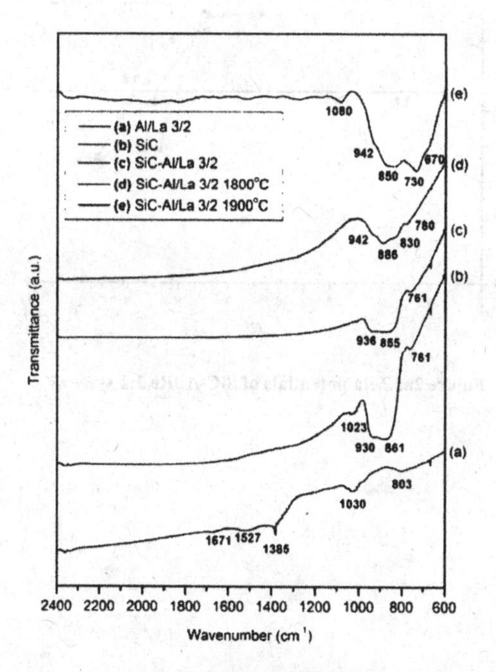

Figure 1: FTIR Spectra of hydroxide precipitates, uncoated SiC, and SiC - Al/La 3/2 before and after sintering at 1800°C and 1900°C

The effect of the coating on the SiC surface on the behavior of the powder in an aqueous suspension was analyzed by measuring the zeta-potential as a function of pH. It is well known that the charge is determined by the composition of the surface layer. The values of the isoelectric point (IEP) of SiC are reported to be in the range from pH 2 to 6[16]. The lower values, similar to silica powders are frequently attributed to the presence of a thin layer of SiO_2. Our results shown in Figure 2 a, b show an isoelectric point at the pH of 3.7, result which is consistent with the FTIR results which identified SiO_2 in our starting SiC material. Figure 2 (a) and (b) shows that the aluminum-rare earth coating significantly changes the behavior of the SiC powder in aqueous suspensions. For all coated powders, the curves illustrating the zeta potential values versus pH, show IEP values shifted towards higher pH values, which are actually characteristic for alumina-based compounds. The zeta potential results display the expected behavior of metal hydroxide precipitates coating the SiC powders as shown by other researchers[17]. The observed shifts are proving that a coating has been achieved.

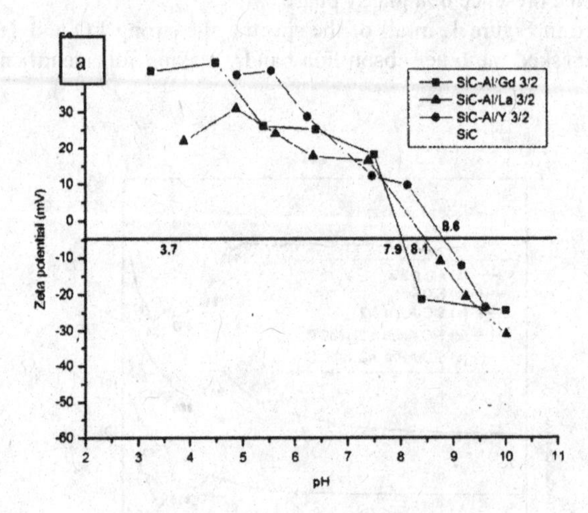

Figure 2a: Zeta potentials of SiC-Al/Re 3:2 systems

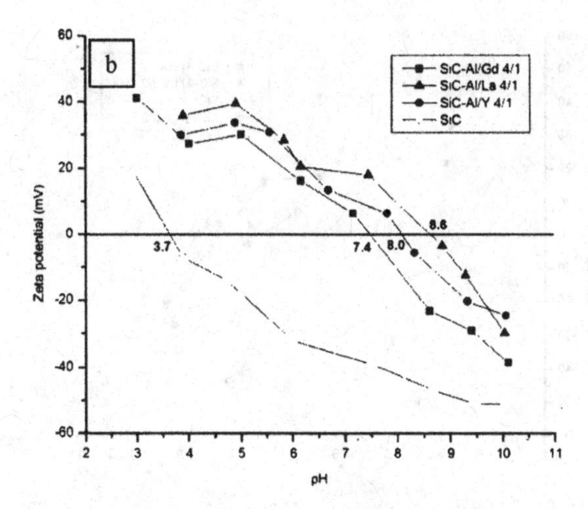

Figure 2b: Zeta potentials of SiC-Al/Re 4:1 systems

Some limited trials were performed with the HX3 surfactant, which is specially designed to coat SiC. It was believed that coprecipitating onto the already coated SiC may lead to better adhesion of the coprecipitating particles. The zeta potential variation versus pH of the coated HX-SiC samples (Figure 3), show also a shift in the IEPs towards higher pH values, but the overall values are lower than for those obtained by coating SiC in the absence of HX. Also, comparing coating of SiC with Al(III) only in the presence of HX led to lower value of the IEP than for the powders coated with mixed Al(III)-La(III) hydroxides. These results cannot be fully explained, since the complete chemical information on the Huntsman HX surfactant is proprietary. The combination of the surfactant and precipitated Al also correlates to previous investigated compositions, using the surfactant and additions of micron sized AlN; the finer particles may further maximize the performance of that composition.

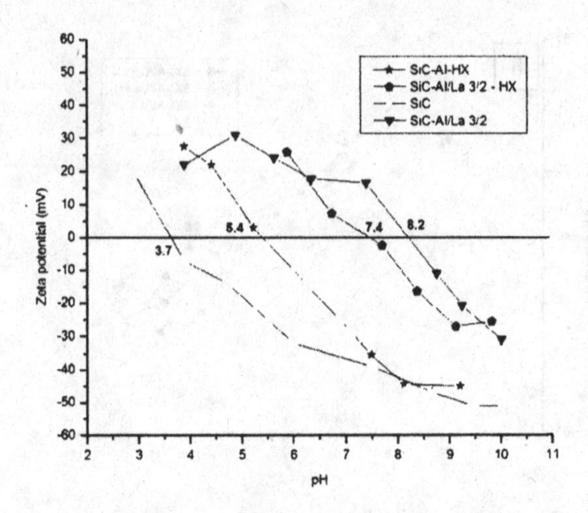

Figure 3: Zeta potentials of SiC-Al/La system with HX3 addition

Scanning electron microscopy was also employed in order to see if any visible changes to the powder surface were noticeable upon coating. SEM image of the uncoated SiC (Figure 4) shows areas with small particles (150-400 nm) of already identified SiO_2 along with larger polydisperse SiC particles (600 nm – 1.7 μm). Rare-earth coated SiC particles in the presence of HX (Figure 6) and without it (Figure 5) show a larger number of small particles having a narrower size range (350-400 nm) deposited on SiC. These images show that the coprecipitation allows obtaining coatings consisting of particles with a narrow size range distribution.

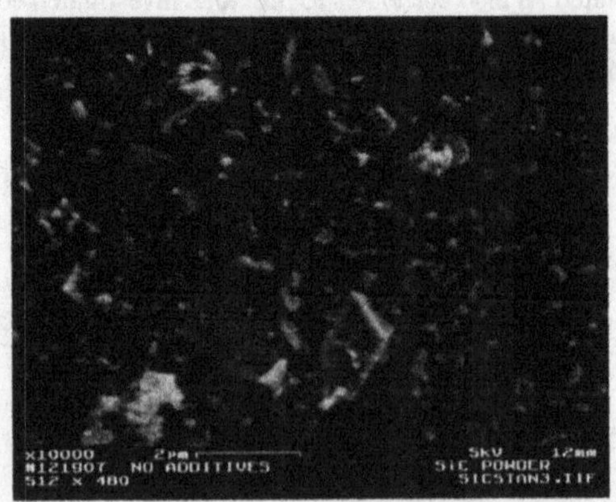

Figure 4: Uncoated SiC powder

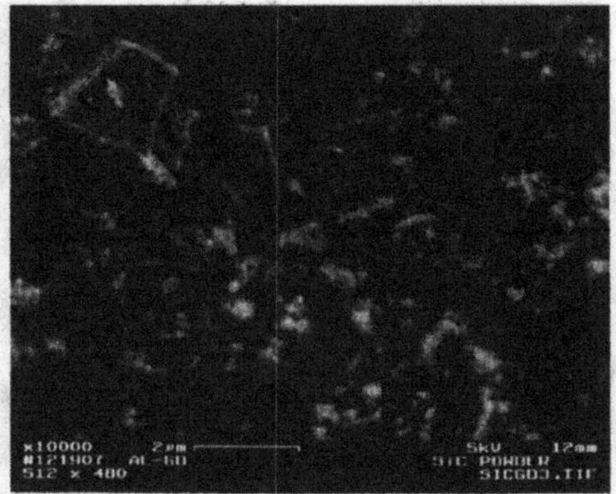

Figure 5: SiC coated with Al and Gd

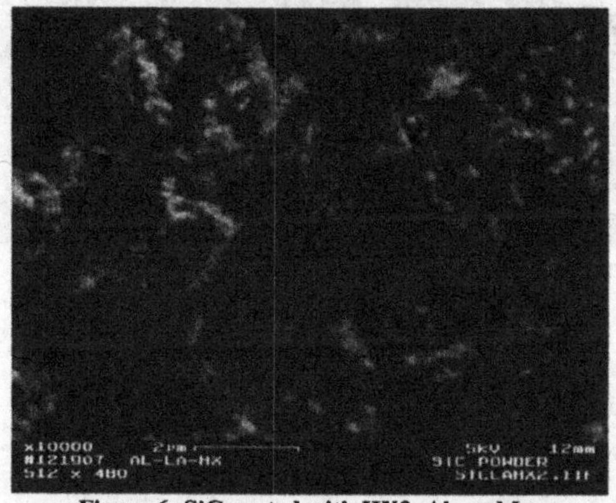

Figure 6: SiC coated with HX3, Al, and La

Transmission electron microscopy suggests the presence of intergranular films, but more work is needed in obtaining better microscopy. The process of crushing sintered samples into powders, although simple and quicker than solid sample preparation for TEM analysis, resulted in intact grain boundaries being difficult to find. Some small regions of connected grains, shown in Figure 7, appear to show the presence of an amorphous film, and Figure 8 shows what is believed to be a film between another two grains. Improved sample preparation and microscopy, as well as the use of characterization tools like EELS and EDX should allow for better characterization of the grain boundary regions in these samples.

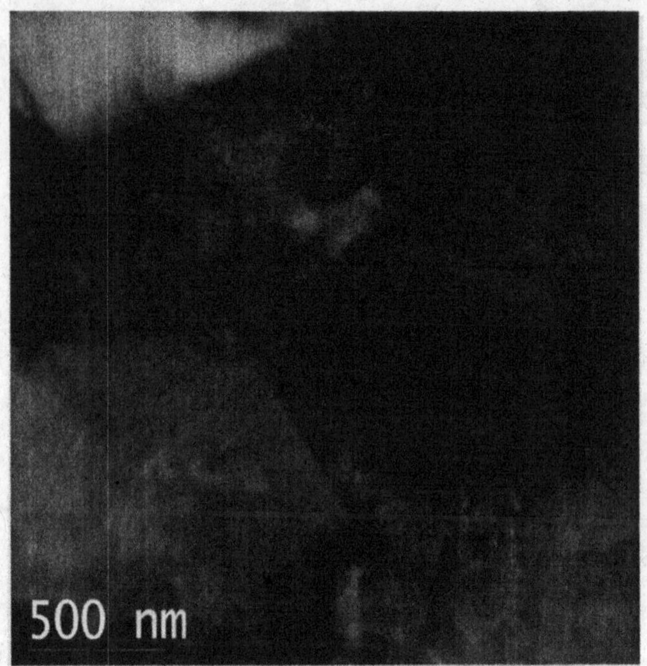

Figure 7: TEM of amorphous region between 2 grains

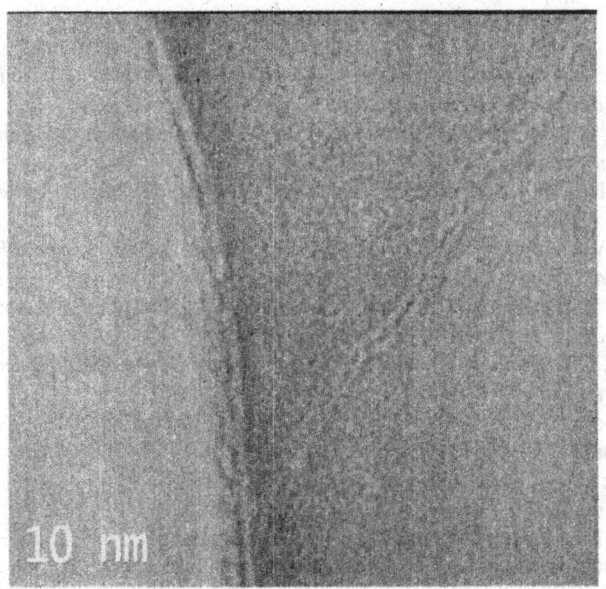

Figure 8: TEM of intergranular film

Conclusions:

Coprecipitation has been developed as a viable method for efficiently coating SiC powders with aluminum and rare earth sintering aids. After coating, zeta potential measurements showed that the electrokinetic properties of SiC powder in aqueous medium have been modified. The structure of the coating layer was evaluated by Fourier transformed Infrared spectroscopy (FTIR). FTIR spectroscopy identified a silica-aluminum-rare earth glass structure after annealing of the coated SiC samples at 1900°C. The size and distribution of the coating particles was investigated by scanning electron microscopy (SEM). Scanning electron microscopy (SEM) images showed that coprecipitation led to a coating consisting of small (350-400 nm) particles in a narrow size range. Preliminary TEM examination confirms the presence of grain boundary phase in sintered samples.

References:

1 M. Bakas, V.A. Greenhut, D.E. Niesz; J. Adams, and J. McCauley. *Ceramic Engineering and Science Proceedings* **24** no3 351-8 2003.
2. P.F. Becher, G.S. Painter N. Shibata, R.L. Satet , M.J. Hoffmann, S.J. Pennycook. *Materials Science and Engineering A* 422 (2006) 85–91
3 Y. Zhou, K. Hirao, M. Toriyama, Y. Yamauchi, and S. Kanzaki. *J. Am. Ceram. Soc.*, **84** [7] 1642–44 (2001).
4 C. Ziccardi, V. Demirbas, R. Haber, D. Niesz, and J. McCauley, *Ceramic Engineering and Science Proceedings, Vol. 26, No.4,* (2005)
5 S. Mercurio, R. Haber, *Advances in Ceramic Armor III: Ceramic Engineering and Science Proceedings, Volume 28, Issue 5* pg 155 (2007).
6 J. Yang, F. J. Oliveira, R. F. Silva and J. M. F. Ferreira. *Journal of the European Ceramic Society* 19 (1999) 433±439
7 Maria P. Albano, Liliana B. Garrido. *Ceramics International* 29 (2003) 829–836
8 D. Sciti, A Balbo, A Bellosi. *Advaned Engineering Materials* 2005, 7, No. 3
9 S.A. Alekseev, V.N. Zaitsev, *Chem. Mater* 2007, 19, 2189.
10 A.R. Oliveira, M.N.P. Carreno, *J. Non-Cryst. Solids* 2006, 352, 1392.
11 S. Daniele, L.G. Hubert-Pfalzgraf, *J. Sol Gel Sci & Technol.* 2005, 35, 57.
12 K. Nakamoto, Infrared and Raman spectra of inorganic and coordination compounds, *5th Ed.* 1997
13 M. Micusic, M. Omastova, K. Boukerma, A. Albony, M. Chehimi, M. Trchova, P. Fedorko, *Polymer Engn & Sci* 2007, 1198.
14 P. Tarte, Proc. Int. Conf. on Non-Cryst. Solids, Delft, 1964, p.549.
15 J. Marchi, D.S. Morais, J. Schneider, J.C. Bressiani, A.H.A. Bressiani, *J. Non-Cryst. Solids* 2005, 351, 863.
16 S. Novak, J. Kovac, G. Drazic, J.M.F. Ferreira, S. Quaresma, *J. Eur. Ceram. Soc.* 2007, 27, 3545.
17 A. Balbo, D Sciti, A Costa, and A Bellosi., *Materials Chemistry and Physics* 103 (2007) 70–77

THE POSSIBLE ROLES OF STOICHIOMETRY, MICROSTRUCTURE, AND DEFECTS ON THE MECHANICAL BEHAVIOR OF BORON CARBIDE

Ryan McCuiston*, Jerry LaSalvia and James McCauley
U.S. Army Research Laboratory
AMSRD-ARL-WM-MD
Aberdeen Proving Ground, MD 21005

William Mayo
Rutgers University, Department of Materials Science and Engineering
607 Taylor Road
Piscataway, NJ 08854

ABSTRACT

The ultimate failure of boron carbide in a ballistic event depends on a number of elastic and inelastic processes whose length scales vary from the atomistic to macroscopic, including the various deformation modes, damage nucleation and accumulation processes, from the atomic structure level through the micro-structural to the macro-structural level. The expanded use of boron carbide as an armor ceramic is dependant upon understanding these mechanisms and using this knowledge to design and produce improved boron carbide. In recent years, ARL, in collaboration with other researchers, has identified previously unknown inelastic phenomena in boron carbide including solid-state amorphization at the nano-scale and subsequent shear localization at the microstructural level. Recent characterization of boron carbide powders has raised questions concerning the presence of $B_{13}C_2$, the stoichiometry of B_4C, and the accuracy of the commonly accepted phase diagrams. In addition, the characterization of commercial boron carbide ceramics has revealed the presence of a very wide range of microstructural defects. These results have led to density functional theory (DFT) calculations on the stability of various boron carbide polymorphs, high-pressure phase transition studies, and raised questions concerning the influence of stoichiometry and microstructural defects on quasi-static and dynamic mechanical properties. A review of recent findings and possible interpretations will be discussed.

INTRODUCTION

As a technical ceramic, boron carbide has found uses in many severe applications. It is used in ceramic composite forms for cutting tools, in grit blast nozzles, as wear coatings, and in the B^{10} isotope enriched form for nuclear shielding and control rods.[1-3] More recently, boron carbide has received attention for use as a thermoelectric material due to its high Seebeck coefficient and in beta voltaic batteries due to its tolerance for radiation damage through self healing.[4]

As an armor ceramic, boron carbide is attractive due to its high hardness, high elastic modulus, high Hugoniot elastic limit and, of great importance for light weight armor applications, its low density. Boron carbide does have some detractions, including the high cost associated with hot pressing and finish machining, as well as low fracture toughness, which affects both handling durability and dynamic performance. It is known from plate impact experiments that boron carbide also suffers an apparent loss of shear strength in the range of 20

---

* Work performed with support by an appointment to the Research Participation Program at the U.S. ARL administered by the Oak Ridge Associated Universities.

to 23 GPa, but the reason(s) for the loss remained unexplained. Two newly recognized damage mechanisms in boron carbide, solid state amorphization[5-9] and shear localization[10] may play a role in the ballistic response of boron carbide, particularly the loss of shear strength. Solid state amorphization has been observed in static indents[6], dynamic indents[8] and on ballistic fragments[5], which can lead to the erroneous belief that it is the dominant mechanism in determining the ballistic behavior of boron carbide. However, there is a degree of system dependence on the activation of these mechanisms in a ballistic event and the observed loss of shear strength potentially attributed to them, but this has not yet been fully understood.

Taking a more global view of boron carbide, there are other factors that may also influence the ballistic response, including but not limited to, phase impurities in the starting powder, boron to carbon stoichiometry differences and microstructural or processing defects. These issues must also be examined and dealt with if boron carbide is to be improved and utilized in armor against a wider variety of threats. This paper will introduce and discuss the aforementioned issues of powder impurities, boron-carbon stoichiometry and microstructural defects as they may relate to the mechanical behavior of boron carbide.

POWDER IMPURITIES

The following description of boron carbide powder production is summarized from Wilson and Guichelaar.[11] The majority of industrially manufactured boron carbide powder is produced in electric arc furnaces via the carbothermal reaction,

$$2B_2O_3 + 7C = B_4C + 6CO \tag{1}$$

The reaction is endothermic until ~1550°C and is usually carried about slightly above the melting point of B_4C, ~2450°C. The starting boron oxide (B_2O_3), often referred to as anhydrous boric acid, is obtainable in high purity. The hydrated form, boric acid (H_3BO_3), may also be used, but complex reactions that arise as a result of the presence of water vapor make this unattractive. The carbon source can be of many forms though usually graphite turnings or pet (petroleum) coke is used. The particle sizes of both the boron oxide and carbon are usually fairly coarse, several millimeters, to keep dust loss down. The B_2O_3 and C are batched together in a mixer, typically with the addition of some recycled mix from previous runs, called revert. The batch is usually made boron rich to account for boron oxide vaporization. This makes carbon the limiting reactant. (Of note is the large amount of carbon monoxide (CO) that is generated by the reaction. CO is both flammable and poisonous and must be carried away from the furnace area, typically using large ventilation hoods.)

Figure 1 (left) shows a cross sectional diagram of a typical AC arc furnace used in the production of boron carbide. The bottom of the furnace is first built up with pet coke and then a layer of revert mix. Next a layer of graphite chips is laid down and the electric arc is struck. A small amount of the new mix is added until the arc is stabilized. The mix is then added slowly by hand as the graphite electrodes are retracted. For a 2 meter diameter furnace capable of holding ~7000 kg of mix, the production run will take 18 to 20 hours. Once the run is complete, an ingot of boron carbide, ~1000 kg, as shown in is Figure 1 (right), is extracted from the unreacted mix, ~5000 kg. (The 5000 kg of unreacted mix will be used in the next run as revert) The balance of ~1000 kg is typically dust loss due to the ventilation, though it is collected in filter assemblies.

The ingot is broken into large chunks which are classified by hand into various grades depending on the color, density and crystal texture. The chunks are then crushed and milled to a designated powder size for that grade. The powder may go through several purification steps, depending on grade, to remove excess B_2O_3 and carbon, as well as impurities such as iron that are introduced by wear of the comminution equipment. Sulfuric and nitric acid washes are a common purification step.

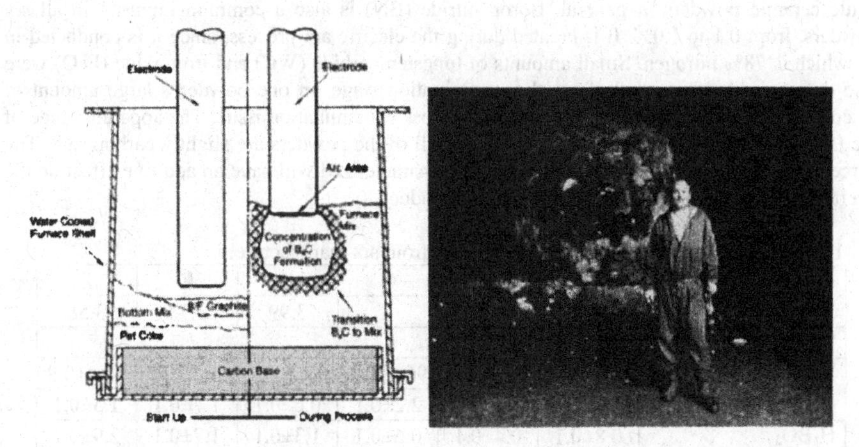

Figure 1. (left) Electric arc furnace used in the production of boron carbide. From Ref 11. (right) Boron carbide ingot produced in the electric arc furnace. Photo from K. Schwetz, ESK.

This direct quote from Wilson and Guichelaar gets to the heart of electric arc processed boron carbide.

"*Arc furnace processing of non-oxide ceramics is economical but limited to products that can tolerate a variable final chemistry, course grains and particle shapes that result from crushing processes.*"

The production of boron carbide by the electric arc process has obvious limitations to how controlled the final product can be. There are several areas in which variability could be introduced, including mixing of the batch, the human intervention required for feeding the furnace and separating the ingot and the purification of the powder after comminution. To help illustrate this variability, boron carbide powders from six different manufactures were obtained for comparison purposes. It is assumed that all six powders are produced using the electric arc process. The powders represent both domestic (USA) and foreign powder producers (Europe and Asia). The majority of the powders obtained are suitable for hot-pressing as-is. The powders were examined for phase composition by X-ray diffraction (XRD) using a full pattern fit Rietveld analysis available with JADE. X-ray fluorescence (XRF) was also conducted to determine the elements present to aid in the phase determination.

Table I is a summary of the detected phases and some pertinent properties as provided by the manufacturers. Carbon, in the form of graphite, and boric acid (H_3BO_3) were the main impurities detected in all six of the powders. The graphite content ranged from 0.6 to 2.2% while the boric acid ranged from 0.4 to 2.9%. The large majority of unreacted carbon and boron oxide from the manufacturing is washed away during the purification steps, so it is likely that the presence of free carbon is most likely due to intragranular graphite precipitates within the boron carbide powder being exposed during milling. This type of carbon is difficult to remove. The boric acid is from the oxidation and hydration of the surface of the powder particles, as occurs with all non-oxide ceramic powders in general. Boron nitride (BN) is also a common impurity in all six powders, from 0.1 to 1.0%. It is created during the electric arc process, since it is conducted in air which is 78% nitrogen. Small amounts of tungsten carbide (WC) and iron oxide (FeO) were also detected, likely residuals from the comminution stage. In one powder a large amount of silicon carbide was detected, which could be a cross contamination issue. The apparent range of the B/C ratio is from 4 to 3.58, which indicates all of the powders are slightly carbon rich. The perceived slight differences among the powders characterized will have an additive effect on the overall properties of the finished boron carbide product.

Table I. Comparison of boron carbide powder from six manufacturers.

Manufacturer	**A**	**B**	**C**	**D**	**E**	**F**
B:C Ratio [1]	3.9	3.78	4	3.99	3.73	3.58
Phase Content [2]						
B_4C / $B_{13}C_2$ [3]	97.6±0.9	97.5±1.0	96.6±0.7	97.2±0.7	96.2±0.7	95.5±0.9
C (graphite)	1.2±0.1	1.1±0.1	2.2±0.1	0.6±0.1	1.7±0.1	1.3±0.1
H_3BO_3	0.8±0.1	0.4±0.1	0.6±0.1	1.3±0.1	1.7±0.1	2.9±0.1
BN (Hex)	0.4±0.1	1.0±0.1	0.5±0.1	0.1±0.1	0.2±0.1	0.3±0.1
WC	n/a	0.1±0.1	< 0.1	n/a	n/a	< 0.1
SiC	n/a	n/a	0.1±0.1	0.9±0.1	0.3±0.1	0.1±0.1
FeO	n/a	0.1±0.1	n/a	n/a	n/a	n/a
Powder Characteristics [1]						
Avg Grain Size (μm)	0.8	0.84	< 2.0	0.82	2.5	--
Surface Area (m^2/g)	--	19.47	--	16.7	3.46	--

1. As reported by the manufacture. 2. Measured using XRD and XRF with Rietveld analysis. 3. Still under review

BORON-CARBON STOICHIOMETRY

Boron carbide has a wide compositional range, currently established from $B_{4.3}C$ to $B_{11.2}C$, 18.8 to 8 mol.% carbon. The commonly available boron carbide powders and monolithic products tend towards the carbon rich limit of boron carbide and are approximately B_4C as shown in Table I. The use of "B_4C" as shorthand for boron carbide is a bit of a misnomer because to achieve that, excess carbon is added which results in a boron carbide – carbon composite.

As was shown in the phase diagram of Samsonov[12] 50 years ago, Figure 2 (left), boron carbide appeared to have two distinct phases within a wide compositional range, $B_{13}C_2$ and $B_{12}C_3$ (B_4C). These phases were established due to the observed microstructural differences, XRD patterns and variations in measured properties of density, micro-hardness and electrical resistivity, as shown in Figure 2 (right). Over time the phase diagram was refined to show a single wide compositional field, but there exists property variations within it.

Recently Chheda et al.[13], in an effort to increase the fracture toughness, produced a series of boron-rich boron carbides from B_4C to a composition that approached $B_{13}C_2$. It was found that

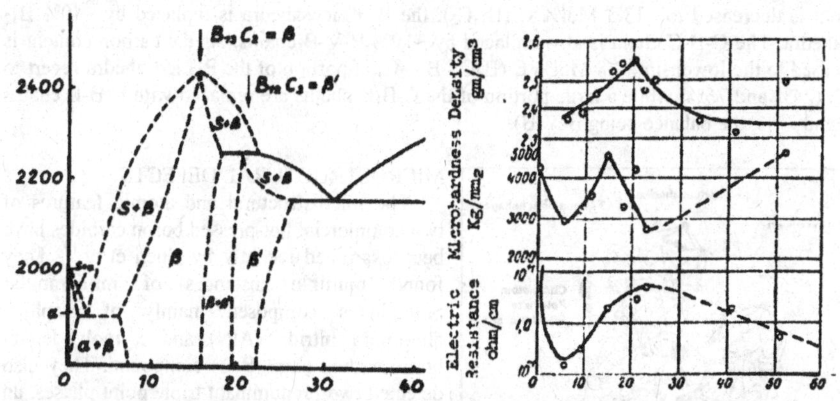

Figure 2. (left) Phase diagram of Samsonov in °C and wt.% C. (right) Data of Samsonov[12].

the fracture toughness increased from ~3 $MPa.m^{1/2}$ for B_4C, to ~5 to 5.5 $MPa.m^{1/2}$ as the composition neared $B_{13}C_2$. It was also shown that the c lattice parameter of the unit cell increased as a function of boron content, but the increase halted near the composition $B_{13}C_2$. Most recently, Tkachenko et al.[14] studied the fracture strength of B_4C and a composite of $B_{13}C_2$ + 30 wt.% B_4C, both hot pressed with identical additions of Al and Al_2O_3. They found the fracture strength of $B_{13}C_2$ + B_4C composite to be significantly higher than pure B_4C, but the reason for this was not described beyond that it could be due enhanced growth twins in the grains.

The inherent physical, mechanical and electrical properties of boron carbide change with composition due to the unique atomic structure of boron carbide and most boron-rich solids in general. The unit cell of boron carbide is comprised of twelve atom icosahedra bonded by three atom chains, as shown in Figure 3 (top). On an absolute number basis, the icosahedra are predominately composed of boron and are quite deformable under stress, while the chains have more carbon and are quite stiff. The chain is largely responsible for the performance of boron carbide and thus changing the chain composition and distribution of chain types, changes the performance.

The elemental makeup of the icosahedra and chain, as a function of composition, is a hotly debated topic.[4, 15-18] It appears that the only thing that can be agreed upon is that the unit cell is composed of icosahedra and chains. There is postulated to be chains of B-B-B, C-C-C, C-B-C, C-B-B, and B-V-B (V= vacancy) and icosahedra of B_{12}, $B_{11}C$, and $B_{10}C_2$, where even then the C can be located either on polar or equatorial sites. The distribution of these multiple chains and icosahedra varies with composition and there appears to be a preference for altering the chain or the icosahedra type between a middle composition of $B_{13}C_2$ and the phase boundaries $B_{4.3}C$ and $B_{11.2}C$. This explains why the properties of boron carbide vary significantly between B_4C and $B_{13}C_2$, the unit cell and thus fundamentally the atomic bonding, is changing.

One of the most recently published descriptions of the distribution of chain and icosahedra, as a function of composition, is given by Werheit et al.[18], see Figure 3 (bottom). (The entirety of this description is not agreed upon by the boron carbide community by any means, but is given to emphasize the atomic structural complexity of boron carbide) At the carbon rich limit of boron carbide, ~18.8 Mol% C ($B_{4.3}C$), the structure is composed of $B_{11}C$ icosahedra bonded predominately by C-B-C chains (81%) with a fraction of C-B-B chains (19%). As the carbon content is decreased to ~13.3 Mol% C ($B_{13}C_2$), the $B_{11}C$ icosahedra is replaced by ~40% B_{12} icosahedra. The C-B-C chain is also replaced by ~19% B-V-B chains. As the carbon content is decreased to the lower limit ~8 Mol% C (B_8C - $B_{11.2}C$), a portion of the B_{12} icosahedra revert to $B_{11}C$, (24% and 76%) while a large portion of the C-B-C chains are replaced with C-B-B chains (7% and 77%, the balance being B-V-B).

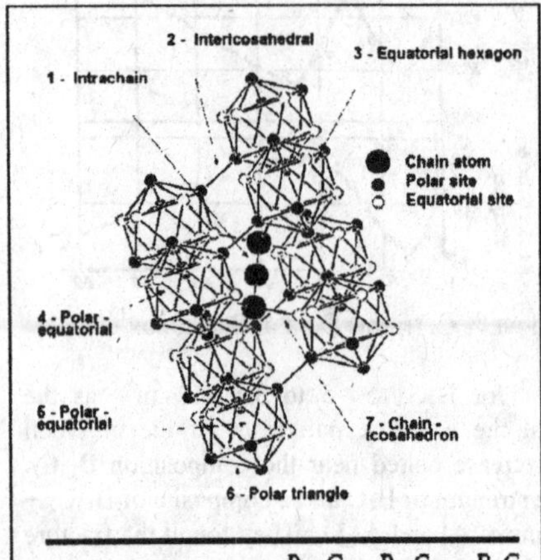

	$B_{4.3}C$	$B_{13}C_2$	B_8C
B_{12} icosahedra	0	42	26
$B_{11}C$ icosahedra	100	58	74
C–B–C	81	62	7
C–B–B	19	19	77
B–□–B (□ vacancy)	0	19	16

Figure 3. (Top) Unit cell of boron carbide. From Ref 16. (Bottom) Distribution of icosahedra and chain composition for $B_{4.3}C$, $B_{13}C_2$ and B_8C. From Ref 18.

MICROSTRUCTURAL DEFECTS

The microstructural and atomic features of two commercial hot-pressed boron carbides have been examined recently by Chen et al.[19]. They found multiple instances of intragranular precipitates composed mainly of graphite, aluminum nitride (AlN) and a molybdenum boron carbon (Mo_2 B-C) compound. They also detected two predominant triple point phases, an iron boride (Fe_2B_{103}) and titanium boride (Ti_3B_4) and intergranular graphite. One of the key findings was the lack of a continuous grain boundary film (intergranular film). The lack of an intergranular film explains boron carbide's high hardness and elastic modulus, as well as its limited fracture toughness.

In a ballistic event, in addition to the size of the defects, their spatial distribution within the bulk is also important. Commercially hot-pressed boron carbide tiles, examined by Vogler et al.[20] using ultrasonic measurements, were found to have spatial variations in the longitudinal and shear sound velocities. When they conducted plate impact tests on the tiles, oscillations in the shock wave, visible with a line VISAR, were determined to have a characteristic length on the order of 50 to 100 µm. They attributed the variation to elastic anisotropy of the boron carbide.

A more thorough examination of the bulk features of boron carbide is of importance due to the large volume of material placed under stress in a typical ballistic event. Figure 4 (left) shows a low magnification image of a polished surface of a commercially-available hot-pressed boron carbide. The noticeable bulk feature of this boron carbide is large graphite particles, many of them in the size range of 50 to 100 µm. The largest particle observed was ~250 µm (0.25 mm). Figure 4 (right) shows the layered structure indicative of graphite. The graphite particles are numerous and are well dispersed throughout the bulk. This is in stark contrast to the anomalous defects that were recently described in hot-pressed silicon carbide.[21] Due to their size and even dispersion they likely originate from processing, perhaps as the remnant of a forming aid.

To study the influence of these large graphite particles on static mechanical properties, a tile of commercially-available hot-pressed boron carbide, the same as observed in Figure 4, was machined into size B bend bars for measurement of the four point flexure strength. 38 bars were tested on a Zwick/Roell Z030 MTM, following ASTM C 1161-02c. The average flexure strength, $S_{(4,40)}$, was 370±40 MPa. The fracture surfaces of several of the weakest and strongest bars were examined. The defects responsible for failure initiation in the examined bars were the large graphite particles. Figure 5 (left) shows the failure origin in a bar that had a flexure strength of 250 MPa. The defect is a large graphitic particle, ~100 μm, that happened to contact the tensile surface. Figure 5 (right) shows the failure origin in a bar with a flexure strength of 410 MPa. The defect responsible for failure is also ~100 μm in size, but was within the bulk, ~100 μm below the tensile surface.

Figure 4. (left) Low magnification image of commercial hot pressed boron carbide exhibiting multiple large graphite defects. (right) High magnification image showing the layered graphite structure.

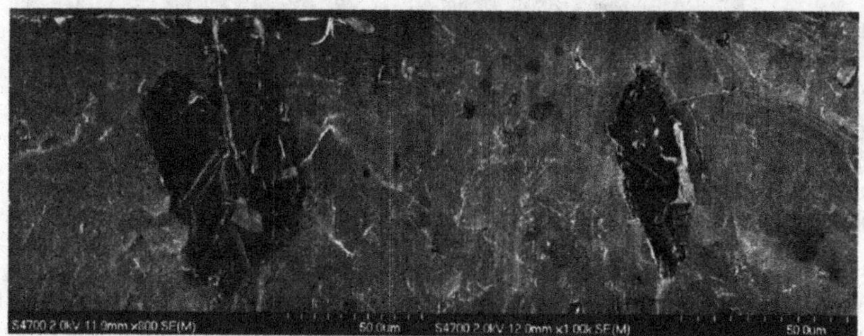

Figure 5. (left) Graphite defect as source of failure in 4pt bend, 250 MPa. (right) Graphite defect as source of failure in 4pt bend, 410 MPa and 100 μm below the surface.

In addition to large graphitic particles. two other defects were observed on the bend bar fracture surfaces. but with less frequency. Figure 6 (left) shows a large boron carbide grain. ~40 μm in size. Due to the elastic anisotropy of boron carbide and the size scaling effect. large grains are much more likely to initiate spontaneous microcracks when under stress. i.e. thermal or mechanical. Figure 6 (right) shows an image of a group of faceted pores. also referred to as negative crystals. While the size of an individual pores is very small. ~1 μm or less. these pores tend to form in large clusters of 20 to 30 μm. typically within a single grain. Combined they may act as a much larger flaw with a low modulus. These clusters are also observed in commercial boron carbide powders, but their formation is not well understood yet.

Figure 6. (left) 40 μm boron carbide grain observed on 4pt bend fracture surface. (right) Negative crystal cluster, 20 μm. observed on 4pt bend fracture surface.

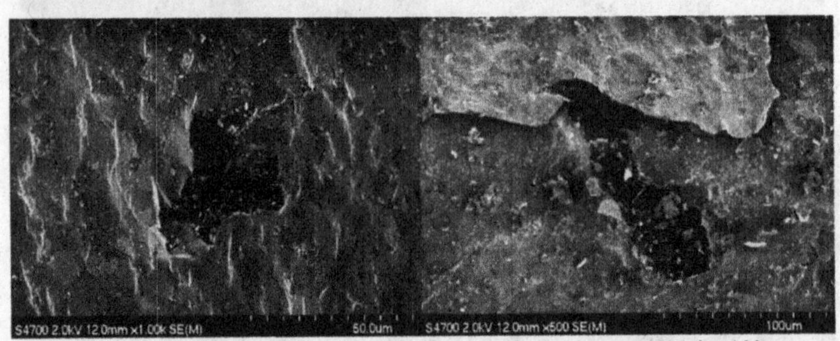

Figure 7. (left) 30 μm graphite defect observed on a near impact fragment. (right) 100 μm graphite defect observed on a far impact fragment.

Several fragments were examined from ballistically tested boron carbide tiles, the same type as used for the bend bars. The fragments were obtained close to the impact site as well as far away. towards the tile edge. The fragments removed from the impact crater were ~3 to 5 mm in size. The fragments further from the impact were typically larger ~10 mm. The fragments were carefully removed from the target and examined in an electron microscope as is. The typical fragment shows a nearly full transgranular fracture surface covered in a fine comminuted powder. Figure 7 (left) shows the largest observed defect in a near fragment. It is a graphite particle, ~30 µm in size. This is smaller than the large graphite particles that were observed. This may be due to the large particles being obliterated or not exposed. It should be pointed out that the fracture path does not appear to have been altered by the presence of the particle. Figure 7 (right) shows the largest defect observed in a far fragment, ~100+ µm in size. Unlike the near fragments, the fracture paths of the far fragments appear to be more influenced by the graphitic particles. This may be due to the different stress state far from the impact.

SUMMARY AND CONCLUSIONS

Boron carbide is attractive as an armor ceramic for its low density and good mechanical properties. It is very effective against a range of threats including small arm, long-rod penetrator, and shape charge threats.[22] However, there are several issues associated with impurities, microstructure and defect variability in bulk boron carbide and current armor-grade variants that may affect its performance. In addition to not yet fully understood issues of solid state amorphization and shear localization, boron carbide may also have issues with powder impurities, boron-carbon stoichiometry and bulk microstructural defects. Boron carbide powders from six manufacturers were characterized for impurity phases. It was found that there are varying levels of free graphite, boric acid and boron nitride in all the powder. There were also smaller levels of comminuted related impurities. These impurities add variability to both the processing and the mechanical properties of the finished product. The phase field of boron carbide is very broad and it is established that the boron to carbon stoichiometry influences the mechanical properties. The optimal stoichiometry is not yet known nor are the methods established for reliably producing them. This is an area ripe for research. A bulk examination of commercially hot-pressed boron carbide was conducted. Large graphite particles are the most numerous defect observed. Additionally, large boron carbide grains and clusters of negative crystals were found. Removal of large graphite defects may be a first step to improve the mechanical properties and reduce the variability of boron carbide for future applications. Finally, bulk boron carbide ceramics are extremely complex materials with varying characteristics among the multiple commercial producers and laboratories and trying to draw broad conclusions based on quasi-static and dynamic mechanical properties, as well as armor performance, must be approached with caution.

ACKNOWLEDGEMENTS

The authors would like to thank Matt Motyka, Herb Miller and Russ Kress for their technical support. The authors would also like to thank Giovanni Fanchini, Munjal Chheda and Mike Normandia for valuable technical discussions.

REFERENCES

1. F. Thevenot, "Boron Carbide - A Comprehensive Review", *J. Eur. Cer. Soc.*, **6**(4) pgs 205-25, (1990)
2. K. Schwetz "Boron Carbide, Boron Nitride and Metal Borides", Ullmann's Encyclopedia of Industrial Chemistry, 6th Ed., **5**, pgs 497-513, (2003)
3. A. Lipp, "Boron Carbide; Production, Properties, Application", *Tech. Rundsch.* **57** (14,28,33), (1965) and **58** (7), (1966)
4. D. Emin, "Unusual properties of icosahedral boron-rich solids", *J. Solid State Chem.*, **179**(9), pgs 2791-98, (2006)

5. M. Chen, J.W. McCauley and K.J. Hemker, "Shock-Induced Localized Amorphization in Boron Carbide", *Science*, **299**(5612), pgs 1563-66, (2003)
6. D. Ge, V. Domnich, T. Juliano, E.A. Stach and Y. Gogotsi, "Structural damage in boron carbide under contact loading", *Acta Materialia*, **52**(13), pgs. 3921–27, (2004)
7. X.Q. Yan, W.J. Li, T. Goto and M.W. Chen, "Raman spectroscopy of pressure-induced amorphous boron carbide", *App. Phys. Let.*, **88**, 131905, (2006)
8. D. Ghosh, G.Subhash, C.H. Lee and Y.K. Yap, "Strain-induced formation of carbon and boron clusters in boron carbide", *App. Phys. Let.*, **91**, 061910, (2007)
9. D. Ghosh, G. Subhash, T.S. Sudarshan, R. Radhakrishnan and X. Gao, "Dynamic Indentation Response of Fine-Grained Boron Carbide", J. Am. Ceram. Soc., **90**(6), pgs. 1850-57, (2007)
10. J.C. LaSalvia, R.C. McCuiston, G. Fanchini, J.W. McCauley, M. Chhowalla, H.T. Miller and D.E. MacKenzie, "Shear Localization in a Sphere-Impacted Armor Grade Boron Carbide", *Proc. 23rd Int. Symp. Ballistics*, pgs 1329-1338, (2007)
11. W.S. Wilson and P.J. Guichelaar, "Electric Arc Furnace Processes" in Carbide, Nitride and Boride Materials Synthesis and Processing, Ed. A. Weimer, pgs 131-36, (1997)
12. G.V. Samsonov, "Physiochemical Properties of Boron – Carbon Alloys", *Fiz. Metal. i Metalloved. Akad. Nauk. SSSR, Ural Filial* **3**, pgs 309-13, (1956)
13. M. Chheda, J. Shih, C. Gump, K. Buechler, A.W. Weimer,"Synthesis and Processing of boron-rich boron carbide", Final Report, Contract No.: DAAD17-01-C-0029, (2005)
14. Yu.G. Tkachenko, D.Z. Yurchenko, V.K. Sul'zhenko, G.S. Oleinik and V.M. Vereshchaka, "Temperature Effect on Bending Strength of Hot-Pressed Boron Carbide Materials", *Powder. Metall. Met. Ceram.*, **46**(5-6), pgs 254-60, (2007)
15. J. Wang, D.S. Marshall, N. Zein and G. Khrenov, "Defect Formation in Boron Carbide – An *ab-initio* Electronic Structure Study", *Mat. Res. Symp. Proc.*, **691**, G.8.6, (2001)
16. R. Lazzari, N. Vast, J. M. Besson, S. Baroni and A. Dal Corso, "Atomic Structure and Vibrational Properties of Icosahedral B_4C Boron Carbide", *Phys. Rev. Let.*, **83**(16), pgs 3230-3 (1999)
17. G. Fanchini, J.W. McCauley and M. Chhowalla, "Behavior of Disordered Boron Carbide under Stress", *Phys. Rev. Let.*, **97**(3), 035502 (2006)
18. H. Werheit, "Are there bipolarons in icosahedral boron-rich solids?", *J. Phys.: Condens. Matter*, **19**, 186207, (2007)
19. M. Chen, J. McCauley, J. LaSalvia and K. Hemker, "Microstructural Characterization of Commercial Hot-Pressed Boron Carbide Ceramics", *J. Amer. Ceram. Soc.*, **88** (7), pgs 1935-1942, (2005).
20. T. Vogler, W.D. Reinhart and L.C. Chhabildas, "Dynamic behavior of boron carbide", *J. App. Phys.*, **95**(8), pgs 4173-83, (2004)
21. M.P. Bakas, V.A. Greenhut, D.E. Niesz, G.D. Quinn, J.W. McCauley, A.A. Wereszczak and J.J. Swab, "Anomalous Defects and Dynamic Failure of Armor Ceramics", *Int. J. Appl. Ceram. Technol.*, **1**(3), pgs 211-18, (2004)
22. M.J. Normandia, "Boron Carbide Ballistic Performance Above the Amorphization Stress", to be published in the Proceedings of the 32nd Int. Conf. Adv. Ceram and Comp., (2008)

A REVIEW OF CERAMICS FOR ARMOR APPLICATIONS

P. G. Karandikar, G. Evans, S. Wong, and M. K. Aghajanian
M Cubed Technologies, Inc.
1 Tralee Industrial Park
Newark, DE 19711

M. Sennett
Natick Soldier RD&E Center
US Army RD&E Command
Natick, MA

ABSTRACT

Over the last decade, the use of ceramics in personnel armor has increased exponentially. In the past, primarily sintered alumina and hot pressed boron carbide were used as the ceramics of choice for personnel armor. In recent years, reaction bonded ceramics (RBSiC and RBB_4C) were developed and are being employed for personnel armor applications. The evolution of ballistic threats on the battlefield will require continuing improvement of armor ceramic performance. To this end, properties of a variety of candidate armor ceramics related to ballistic impact resistance are compared in this work. Also, process-microstructure-property relationships are examined for these ceramics.

INTRODUCTION

A number of criteria must be considered when selecting materials for use in a personnel armor system for protection against ballistic threats. These include the characteristics of the specific threats to be defeated, the allowable volume and weight parameters of the system, and the system cost. Because of the range of design criteria that exist for armor systems, there is no single "best" armor material for all applications. For example, a material that provides adequate protection against specific threats in one system configuration may become inadequate if the permissible system weight is reduced. For this reason, a number of different ceramic materials have been and continue to be used in personnel armor systems.

Personnel armor can be broadly classified into soft armor (e.g. textile based systems using high performance fibers) and hard armor (systems containing metallic or ceramic inserts)[1-4]. Soft armor is adequate for certain low level threats (e.g. NIJ level III or less), while for higher level threats (e.g. NIJ level IV or higher) metallic or ceramic inserts are needed[1-4]. Low density ceramics (with density 2.5-4.0 g/cc) offer higher hardness and less than half the weight (at the same thickness) in comparison with armor steel (density 7.8 g/cc). Table I summarizes the properties of typical ceramics[4-15] used for personnel armor applications. Most of the listed properties for other manufacturer's materials are obtained from material manufacturers' respective datasheets. Hardness (2 kg Knoop), grain size and fracture mode are from measurements made as a part of the present work. Grain size and fracture mode are also corroborated by previously published data.

Table I. Summary of properties of various ceramics for personnel armor application

Material	Designation	Density ρ (g/cc)	Grain Size (μm)	Young's Modulus E (GPa)	Flex Strength σ (MPa)	Fracture Toughness K (MPa-m$^{1/2}$)	Fracture Mode	Hardness (kg/mm^2)	HK 2 kg (kg/mm^2)	Areal Density lb/sf (PSF)*
Al_2O_3	CAP-3	3.90	--	370	379	4-5	--	1440 (HK 1 kg)	1292	20.2
B_4C	Ceralloy-546 4E	2.50	10-15	460	410	2.5	TG	3200 (HV 0.3 kg)	2066	13.0
	Norbide	2.51	10-15	440	425	3.1	TG	2800 (HK 0.1 kg)	1997	13.0
SiC	SiC-N	3.22	2-5	453	486	4.0	IG, TG	--	1905	16.7
	Ceralloy 146-3E	3.20	--	450	634	4.3	--	2300 (HV 0.3 kg)	--	16.6
	Hexoloy	3.13	3-50	410	380	4.6	TG	2800 (HK 0.1 kg)	1924	16.2
	Purebide 5000	3.10	3-50	420	455	--	TG	--	1922	16.1
	SC-DS	3.15	3-50	410	480	3-4	--	2800 (HK 1 kg)	--	16.4
	MCT SSS	3.12	3-50	424	351	4.0	TG	--	1969	16.2
	MCT LPS	3.24	1-3	425	372	5.7	IG	--	1873	16.8
	Ekasic-T	3.25	1-3	453	612	6.4	IG	--	1928	16.8
SiC (RB)	SSC-702	3.02	45	359	260	4.0	TG	1757 (HK 0.5 kg)	--	15.7
	SSC-802	3.03	45	380	260	4.0	TG	--	1332	15.7
	SSC-902	3.12	45	407	260	4.0	TG	--	1536	16.2
SiC/B_4C (RB)	RBBC-751	2.56	45	390	271	5.0	TG +Ductile Si	--	1626	13.3
TiB_2	Ceralloy 225	4.5	--	540	265	5.5	--	--	1849	23.4

Sources: CAP-3, SC DS: CoorsTek; Ceralloy, Ekasic-T: Ceradyne; Norbide, Hexoloy: Saint Gobain; Purbide: Morgan AM&T; SiC-N: Cercom (BAE); SSC, RBBC, BSC, SSS and LPS – M Cubed Technologies (MCT). Properties for other manufacturer's materials are from their respective websites except for 2kg Knoop hardness, grain size and fracture mode. *Areal density (PSF): weight of 12 x 12 x 1 inch panel in pounds. TG – Transgranular fracture, IG – inter granular fracture

Alumina has been widely used in personnel armor systems due to low cost and relatively low density (3.9 - 4.0 g/cc). SiC and B_4C are harder materials with lower density than alumina. However, their cost is significantly higher than that of alumina. As material was added to evolving armor systems in order to defeat more tenacious threats, a significant weight penalty became associated with the use of alumina for personnel armor. As a result, B_4C and SiC have come to be used more commonly over the past decade in personnel armor systems. Sintered and hot pressed varieties of SiC and B_4C were predominant until the development of reaction-bonded SiC and B_4C (RBSC and RBBC) in the late 1990s[7] and their use in personnel armor starting in the year 2000.

RELATIONSHIP OF MATERIAL PROPERTIES TO PERFORMANCE IN ARMOR APPLICATIONS

To date, armor performance has not been successfully correlated to a single material characteristic or static material property[16] due to the dynamic nature of the ballistic event (nano to micro seconds). Thus, a "proof" test of each armor system under set criteria is always required to determine its adequacy to defeat a particular threat. However, several fundamental material properties have been used to rank various ceramics[16] for initial screening purposes. These properties and their commonly understood roles in resisting ballistic impact are listed in Table II.

Table II. Material properties and their role in ballistic performance

Property	**Role /Effect in ballistic performance**
Microstructure Grain size Minor phases Phase transformation or Amorphization (stress induced) Porosity	Affects all properties listed in the left-hand column below
Density	Weight of the armor system
Hardness	Damage to the projectile
Elastic modulus	Stress wave propagation
Strength	Multi hit resistance
Fracture Toughness	Multi hit resistance, field durability
Fracture Mode (inter vs. trans granular)	Energy absorption

A typical armor system for individual protection against small arms threats includes a ceramic tile backed by a fiber reinforced polymer matrix composite[4]. Several processes occur in a ballistic impact event[14-16] (1) projectile impacts the strike face (the ceramic) imparting kinetic energy, (2) a shock wave is generated in the ceramic. Tensile, compressive and shear stresses are generated in the ceramic with their magnitude dependent on the threat, (3) a pattern of circumferential (cone) and radial cracks is generated in the ceramic, (4) the ceramic imparts stress back on the projectile, (5) projectile is deformed, shattered or eroded, (6) the projectile or fragments can continue to penetrate the ceramic, (7) the fragments (projectile and ceramic) generated can enter into the composite backing and are stopped by it, (8) a residual deformation can be created on the back side of the armor system, (9) momentum is transferred to the wearer's body. If the projectile is not defeated by the armor, excessive back face deformation can be caused (resulting in blunt force trauma to the wearer) and in the worst case, the projectile (or fragments) continues to penetrate wearer's body. For successful projectile defeat, the ceramic must be sufficiently robust to significantly deform it or break it into pieces and reduce the kinetic energy sufficiently so the backing can stop it without excessive back face deformation. The projectile thus must be made to spend sufficient time interacting with the surface of the ceramic. The time during which the projectile does not penetrate the ceramic is called "dwell" and the defeat mechanism is called "interface defeat"[16] if the projectile is sufficiently damaged at the ceramic surface so as to preclude further penetration of the armor system.

CERAMIC MANUFACTURING PROCESSES

Various processes are used to make the ceramic materials commonly used in personnel armor. These processes and their advantages and disadvantages are summarized in Table III.

Table III. Manufacturing processes for ceramic armor materials[7,17-23]

Process	**Materials (Properties in Table I)**	**Advantages**	**Disadvantages**
Hot Pressing (HP)	Ceralloy B_4C, Norbide B_4C, Ceralloy SiC, SiC-N, TiB_2	Lower temperature, lowest porosity	Shape limitation
Solid State Sintering (SSS) or pressureless sintering	Hexoloy SiC, Purbide SiC, MCT SSS SiC	No grain boundary phase, low porosity	Higher temperature, grain coarsening
Liquid phase sintering (LPS)	Ekasic-T, MCT LPS SiC	Lower temperature, fine grains, low porosity	Oxide grain boundary phase
Reaction bonding (RB)	Si/SiC, Si/B_4C (MCT RBSC, RBBC)	Low temperature, excellent complex shape capability	Residual silicon

MICROSTRUCTURES

Samples of most types of ceramics listed in Table I were prepared for microstructural analysis via standard grinding, polishing and etching (if needed) techniques. Figures 1 through 5 show micrographs of these materials' microstructures. Microstructural observations of the above ceramics can be summarized as follows: (1) Hot pressed B_4Cs have single phase microstructure with average grains size 10-15 microns and some porosity (few %). These observations compare favorably with previously published data[17]. Some residual graphite may also be present. (2) Liquid phase sintered SiCs have very fine microstructure with 1-2 micron grains due to typically lower sintering temperatures than solid state sintered SiCs, low porosity (1-2%), and a glassy oxide minor phase (5-10%)[18-19]. (3) In solid state sintered SiCs, the grain size can be tailored to be large (>20 micron) or small (3-5 micron) based on sintering conditions[19-21]. Also, 1-3% porosity is observed based on the sintered density reached. Residual carbon or graphite may be present. (4) Hot pressed[22-23] SiCs have higher densities (almost zero porosity) than sintered varieties, and fine microstructure. (5) RB ceramics have larger (10-50 micron) ceramic grains (SiC or B_4C), a silicon minor phase, and no porosity.

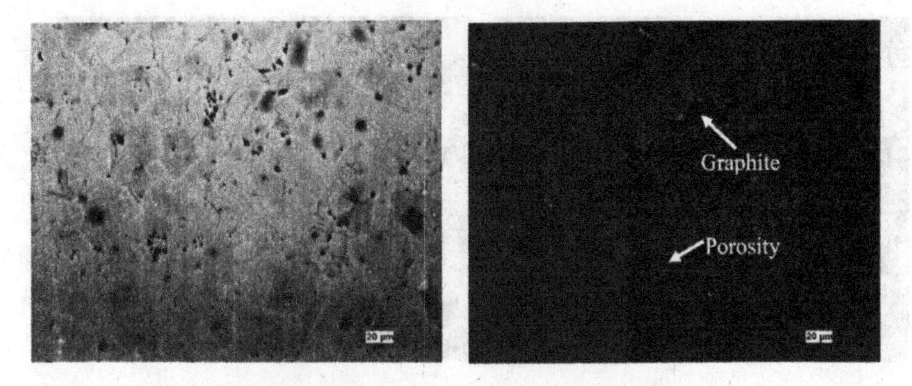

Figure 1. Microstructures hot pressed B_4C ceramics

Figure 2. Microstructure of hot pressed SiC.

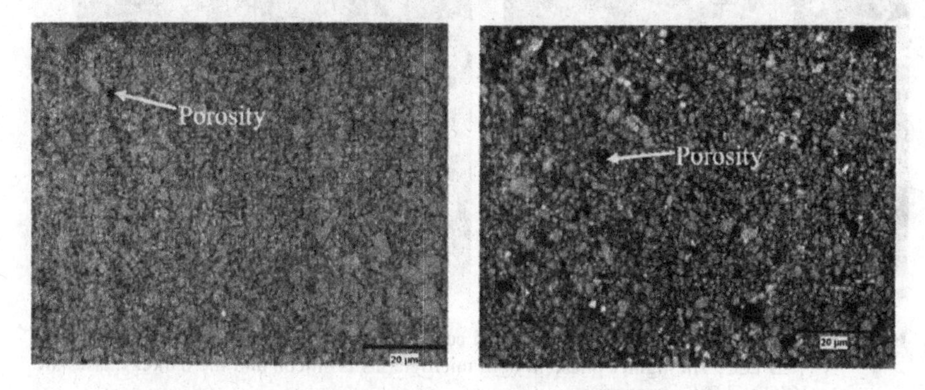

Figure 3. Microstructures of liquid phase sintered SiC ceramics

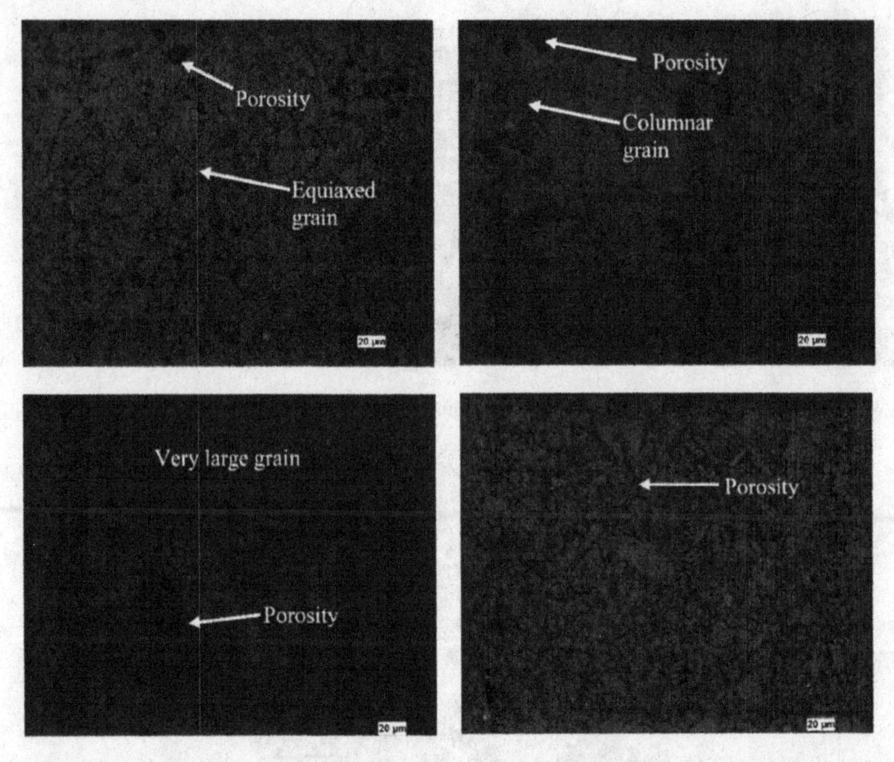

Figure 4. Microstructures of solid state sintered SiC ceramics. Processing conditions can be tailored to produce microstructures with different grain size and grain morphology.

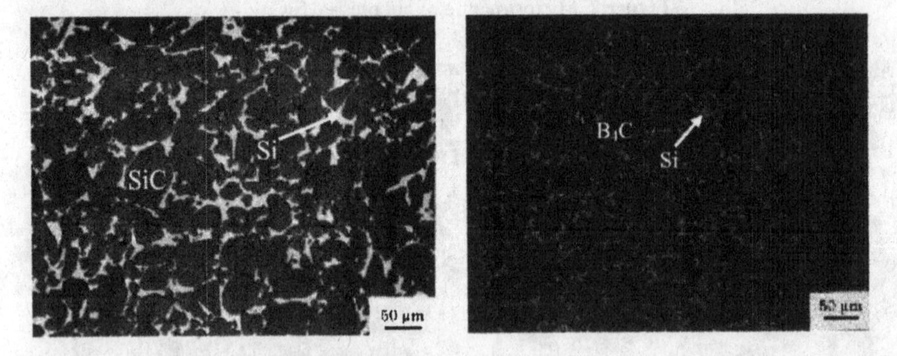

Figure 5. Microstructures of reaction bonded ceramics: **Left:** RB silicon carbide and **Right:** RB B_4C. The lighter phase in both micrographs is silicon and the darker phases are SiC and B_4C, respectively.

HARDNESS AND FRACTURE TOUGHNESS

As a rule-of-thumb, hardness higher than that of the projectile is desirable for armor ceramics. High fracture toughness is also desirable to minimize the shattering of the ceramic on impact which may improve the ability of the material to defeat multiple hits. Unfortunately, these properties tend to follow an inverse relationship in most materials. To illustrate the relationship between hardness and toughness, measurements were carried out on several ceramics and Al/SiC metal matrix composites (MMCs).

Hardness was measured on the Knoop scale with a 2 kg load per ASTM Standard C 1236 using a Shimadzu HMV-2000 hardness tester. For ceramics, the Knoop hardness initially decreases as the applied load increases and then it reaches a stable value. The load that yields a stable hardness is determined to be 2 kg[24]. The Knoop hardness data obtained here compares favorably with that reported in the literature[25]. It must be noted however, that due to the nature of the test, local microstructural effects come into play. For example, the indent size is of the order of 100 micron and the grain size ranges from 1-50 micron. Also, minor phases (e.g. porosity) can interact with the indenter and the length of the indent has to be measured precisely by the operator using a microscope. Therefore, standard deviation in the hardness data can be as much as 5%. For the softer materials tested, a lower Knoop load (0.5 kg) was used (e.g. steels, Al/SiC MMCs etc.). Fracture toughness measurements were carried out using the Chevron Notch method per ASTM 1421.

Figure 6 shows a plot of hardness of several materials as a function of their toughness. Hardness of some representative projectile materials is also shown in the plot. For comparison, the hardness of M Cubed's Al/SiC metal matrix composites (MMCs) with different SiC loadings is also shown. The chart shows an inverse relationship between hardness and toughness. Similar results have been previously reported in the literature[8, 19]. The Al/SiC MMCs have low hardness but very high, metal-like toughness. These materials could be useful as a component of armor systems for defeating low hardness mild steel threats. However, to defeat tool steel and WC projectiles, higher hardness ceramics are typically needed.

EFFECTS OF MINOR PHASES

Due to the nature of their respective manufacturing processes, the candidate armor ceramic materials have a variety of minor phases. For example, hot pressed B_4C and hot pressed SiC have porosity as the minor phase. Solid state sintered SiC also has porosity as the minor phase. Graphite and carbon inclusions are possible in hot pressed B_4C and solid state sintered SiC. Liquid phase sintered SiC has porosity and an oxide glass as minor phases. Reaction bonded materials typically do not have any porosity but have silicon as the minor phase. The effect of these minor phases on the ballistic impact resistance of these materials can be significant. Figure 7 shows a plot of elastic modulus of these materials as a function of minor phase content. In general, the presence of the above listed minor phases leads to a reduction of the elastic modulus. The reduction however, depends on the type of minor phase. Porosity causes the biggest reduction in elastic modulus. Similar effects on other properties critical to ballistic performance can be expected.

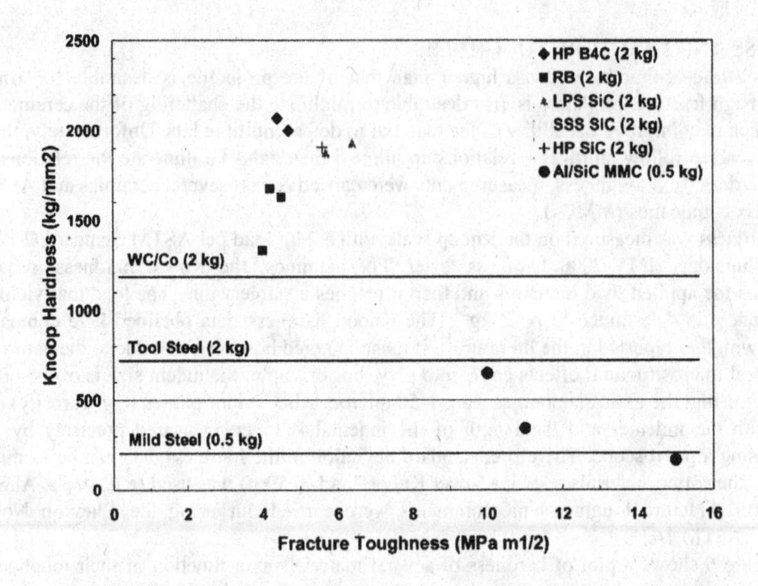

Figure 6. Inverse relationship between hardness and fracture toughness.

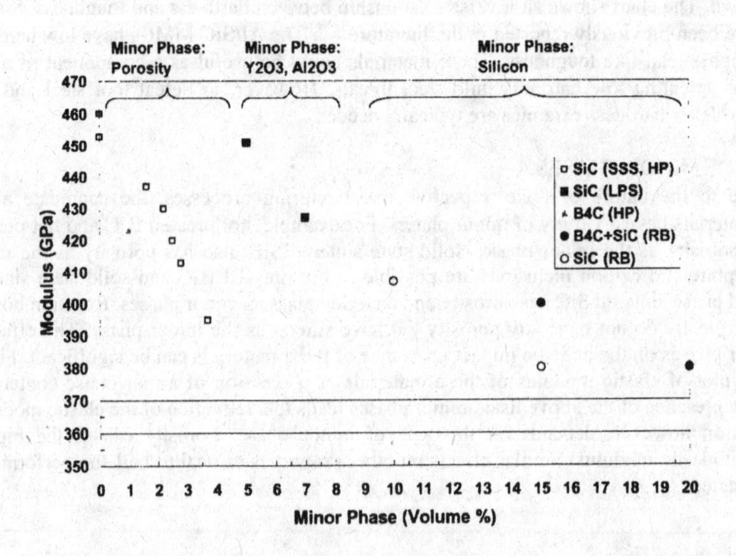

Figure 7. Different minor phases in the ceramics affect modulus differently.

EFFECT OF GRAIN SIZE ON HARDNESS AND STRENGTH

Ceramic properties are also affected by the scale of the microstructure or the grain size. Figure 8 shows the dependence of hardness and flexural strength on the grain size of solid state sintered SiC. Flexural strength in four-point bending was determined following ASTM Standard C 1161. Both hardness and strength decrease as the grain size increases. This is consistent with literature data[26] and the Hall-Petch relationship reported in metals. Thus, microstructure must be controlled carefully to achieve desired ceramic properties and resultant ballistic performance.

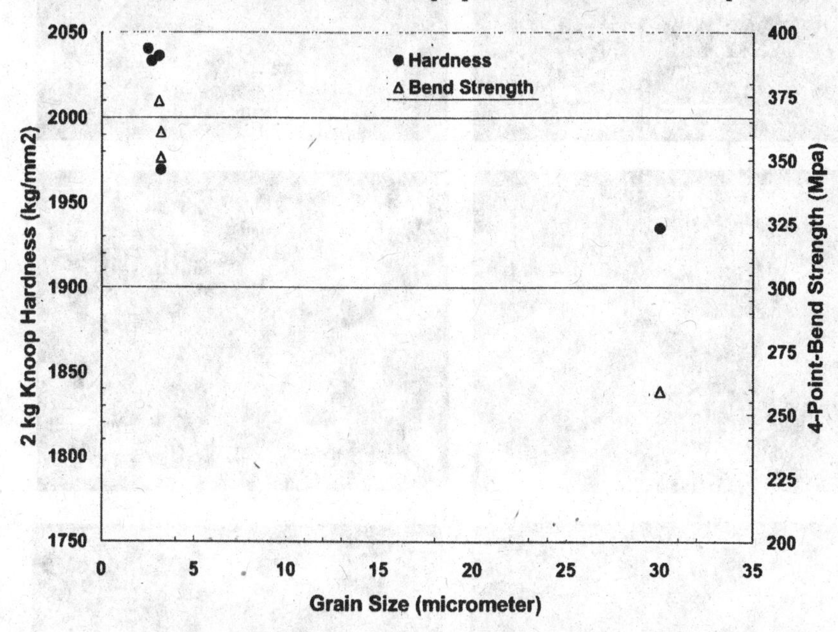

Figure 8. Hardness and strength Vs. grain size relationship in solid state sintered SiC.

FRACTURE MODE

Fracture surfaces of static flexural test bars were analyzed using a scanning electron microscope (JEOL JSM 6480). Figure 9 shows fracture surfaces for various ceramics under quasi-static (low strain rate) failure. In a ballistic event, much higher strain rates exist at the location where the projectile strikes. However, as the distance from the impact location increases, the strain rate decreases. Thus, for regions away from the impact point, quasi-static fracture mode can provide an insight into the failure mechanism. As the strain rate increases, for most materials the failure behavior becomes increasingly more brittle.

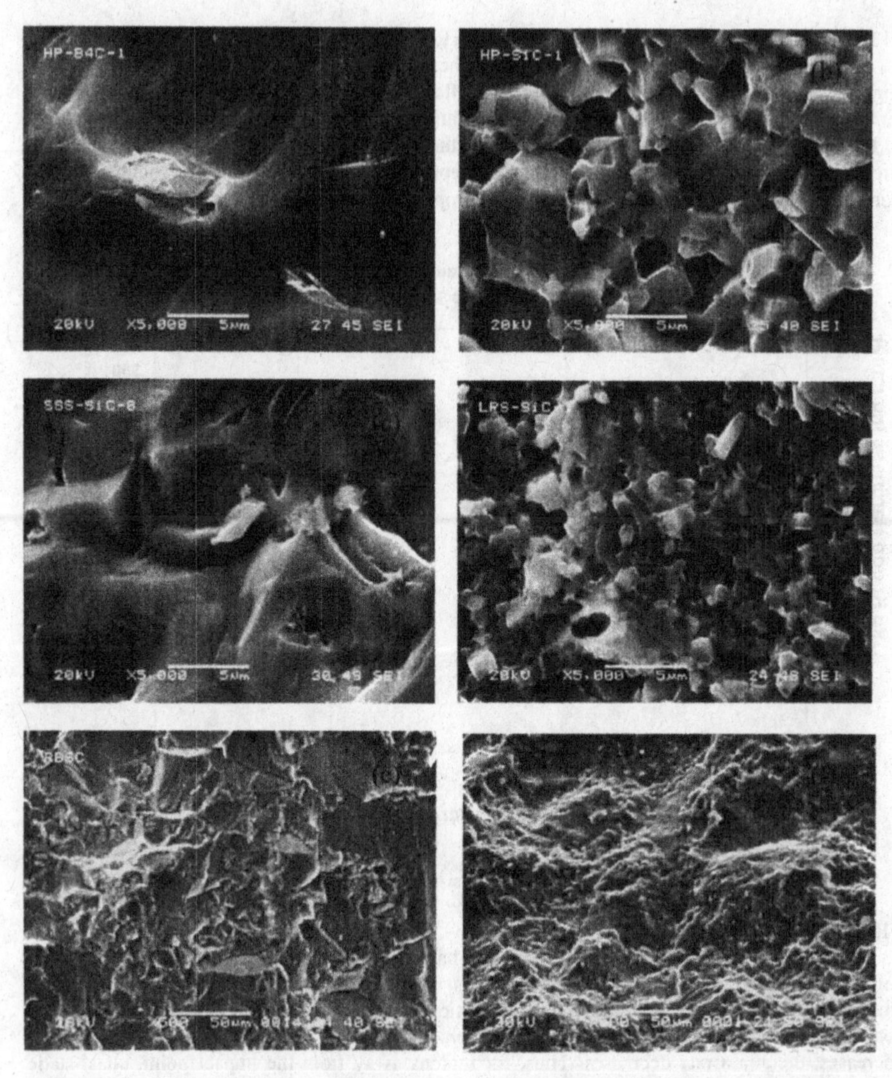

Figure 9. Fracture surfaces of various ceramics (a) hot pressed B_4C – transgranular, (b) hot pressed SiC – intergranular (some trans granular), (c) SSS SiC – Transgranular, (d) LPS SiC – intergranular, (e) RB SiC – transgranular, and (f) RB B_4C – transgranular (with ductile failure of silicon phase).

The hot pressed B_4C and SSS –SiC show transgranular fracture. The hot pressed SiC and LPS SiC on the other hand, show predominantly inter-granular fracture (with some transgranular component). Correspondingly, hot pressed B_4C and SSS –SiC show lower fracture toughnesses than hot pressed SiC and LPS SiC. Thus, inter-granular fracture results in higher toughness. Inter granular fracture follows a more tortuous path through grain boundaries, needing more energy to create a larger fracture surface – resulting in the higher fracture toughness. However, LPS SiC has lower hardness than SSS-SiC. Thus, again, the inverse relationship between hardness and fracture toughness comes into play. Similar interrelationship between hardness, toughness and fracture mode in SiCs has been reported by Flinders et al.[19]. RB B_4C shows ductile-like failure of silicon minor phase whereas silicon is brittle in RB SiC. As a result, RB B_4C shows higher fracture toughness than RB SiC.

AMORPHIZATION AND PHASE TRANSFORMATION

Boron carbide is the lowest density and highest hardness ceramic typically used for personnel armor. However, it has been found recently that its performance against more tenacious threats has been below expectations, especially compared to less hard and denser SiC[9-11]. The recent uncertainties in the performance of B_4C against more tenacious threats have been attributable to shear localization (amorphization) that occurs in this material under high pressure[27]. The maximum contact pressure generated in a ceramic depends on the density, velocity, bulk modulus, and yield strength of the projectile[28]. When this pressure exceeds a critical threshold, amorphization or phase transformation can occur in certain materials. The phase diagram for boron carbide shows that it is not a so-called line compound[29-30]. That is, it does not exist only at the 4:1 atomic ratio of B to C. Rather, B_4C is a continuous phase or solid solution with compositions from $B_{11}C$ to $B_{3.7}C$. The lattice structure of B_4C consists of twelve-atom $B_{12-x}C_x$ icosahedrons connected with three atom intericosahedral chains [29-30]. These chains can be C-B-C, C-B-B or C-C-C as the similarly sized C and B atoms can easily substitute each other. One of these polytypes, B_{12} icoshedra with C-C-C chains has been found to collapse leading to the local amorphization[31] of B_4C at high pressure.

It has also been shown recently that silicon can undergo a pressure-induced phase transformation[32]. This could have adverse effects on the impact resistance of reaction bonded ceramics that have appreciable amounts of residual silicon. Unlike boron carbide, SiC is a line compound and does not undergo amorphization under high applied dynamic pressures. As a result, SiC is becoming increasingly important as a ceramic material of choice against more tenacious projectiles[5-12].

SUMMARY

A variety of ceramics are available for use in personnel armor applications. The key material properties that may be used to guide the selection of ceramics for light armor are density, hardness, grain size, amount of minor phase, phase stability, fracture mode, and toughness. Recently discovered phenomena of pressure-induced amorphization or phase transformation in some materials are becoming increasingly more critical when designing armor systems to defend against high tenacity projectiles.

ACKNOWLEDGEMENTS

This work was partially supported by US DoD SBIR contract #W911QY-06-C-0041.

REFERENCES

[1]www.dsm.com/en_US/html/hpf/dyneema_ud.htm

[2] www.honeywell.com

[3] Ballistic Resistance of Personal Body Armor, NIJ Standard 0101.04) (http://www.ojp.usdoj.gov/nij/topics/technology/body-armor/).

[4] M. van Es, J. Beugels, and J. vanDingenen, "Development of Dyneema/ceramic hybrids for SAPI inserts," Proceedings of PASS 2002, 163-168.

[5] M. Chheda, M. J. Normandia, J. Shih, "Improving ceramic armor performance," Ceramic Industry, January (2006) 124-126.

[6] S. Elliott, "Silicon Carbide ceramic armor," Advanced Materials & Processes, October (2007) 29-33.

[7] M. Aghajanian, B. Morgan, J. Singh, J. Mears and B. Wolffe, "A new family of reaction bonded ceramics for armor applications," in Ceramic Armor Materials by Design, Ceramic Transactions, Vol. 134., J. W. McCauley et al editors, (2002) 527-540.

[8] M. Flinders, D. Ray, R. A. Cutler, "Toughness-hardness trade off in advanced SiC armor," Ceramic Transactions 51 Ceramic Armor and Armor Systems (2003) 37-48.

[9] C. Roberson and P. J. Hazell, "Resistance of different ceramic materials to penetration by a tungsten carbide cored projectile," Ceramic Transactions 51, Ceramic Armor and Armor Systems (2003) 153-163.

[10] C. Roberson and P. J. Hazell, "Resistance of silicon carbide to penetration by a tungsten carbide cored projectile," Ceramic Transactions 51, Ceramic Armor and Armor Systems (2003) 165-174.

[11] N. J. Woolmore, P. J. Hazell, and T. P. Stuart, "An investigation into fragmenting the 14.5 mm BS41 armor piercing round by varying a confined ceramic target set up," Ceramic Transactions 51, Ceramic Armor and Armor Systems (2003) 175-186.

[12] T. M. Lillo, D. W. Bailey, D. A. Laughton, H. S. Chu, and W. M. Harrison, "Development of a pressureless sintered silicon carbide monolith and special shaped silicon carbide whisker reinforced silicon carbide matrix composite for lightweight armor application," Ceramic Transactions 51, Ceramic Armor and Armor Systems (2003) 49-58.

[13] P. J. Hazell, S. E. Donoghue, C. L. Roberson, and P. L. Gotts," The penetration of armor piercing projectiles through reaction bonded ceramics," Ceramic Engineering and Science Proceedings (2006) 143-150 (Proceedings of the 30th International Conference on Composites and Advanced Ceramics, Cocoa Beach, FL, January 2006.

[14] B. James, "Practical issues in ceramic armor design," Ceramic Transactions Vol. 134, Ceramic Armor Materials by Design (2002) 33-44.

[15] M. J. Normandia and W. Gooch, "An overview of ballistic testing methods of ceramic materials," Ceramic Transactions Vol. 134, Ceramic Armor Materials by Design (2002) 113-138.

[16] J. C. LaSalvia, "Recent progress on the influence of microstructure and mechanical properties on ballistic performance," Ceramic Transactions Vol. 134, Ceramic Armor Materials by Design (2002) 557-570.

[17] R. Angers and M. Beauvy, "Hot pressing of boron carbide," Ceramics International, Vol. 10 [2] (1983) 49-55.

[18] K. Y. Chia, W. D. G. Boecker, R. S. Storm, "Silicon carbide bodies having high toughness and fracture resistance and method of making same," U. S. Patent 5,298,470 (1994).

[19] M. Flinders, D. Ray, A. Anderson, and R. A. Cutler, "High toughness silicon carbide as armor, J. Am. Ceram. Soc. 88[8] (2005) 2217-2226.

[20] S. Prochazka, "Silicon carbide sintered body," US Patent 4,041,117 (1977).

[21]J. A. Coppola, N. Hailey, and C. H. McMurtry, "Sintered alpha silicon carbide ceramic body having equiaxed microstructure," US Patent 4,179,299 (1979).
[22]S. Prochazka, "Hot pressed silicon carbide," US Patent 3,853,566 (1974).
[23]A. Ezis, "Monolithic, fully dense silicon carbide material, method of manufacturing and end uses," US Patent 5,372,978 (1994).
[24]J. J. Swab, "Recommendations for Determining Hardness of Armor Ceramics", Int. J. Appl. Ceram. Tech., 1 [3] 219-25 (2004).
[25]A. Wereszczak, H. Lin, and G. A. Gilde, "The effect of grain growth on hardness in hot pressed silicon carbides," J. Mater. Sci. 41 (2006) 4996-5000
[26]R. W. Rice C. C. Wu, and F. Borchelt, "Hardness grain size relations in ceramics," J. Am. Ceram. Soc. 77 No. 4 (1994) 2539-2553.
[27]M. Chen, J. W. McCauley, and K. J. Hemker, "Shock induced localized amorphization in boron carbide," Science Vol. 299 [7] 2003 1563-1566.
[28]P. Lundberg, R. Renstrom, and L. Westerling, "Transition between interface defeat and penetration for a given combination of projectile and ceramic material," Ceramic Transactions Vol. 134, Ceramic Armor Materials by Design (2002) 173-181.
[29]D. Emin, "Structure and single phase regime of boron carbides," Physical Review B, Vol. 38 [9] (1988) 6041-6055.
[30]F. Thevenot, "Boron carbide – a comprehensive review," Journal of the European Ceramic Society, 6 (1990) 205-225.
[31]G. Fanchini, J. W. McCauley, and M. Chowalla, "Behavior of disordered boron carbide under stress," Physical Review Letters, 97 (2006) 035502
[32]V. Dominich and Y. Gogotsi, "Phase transformation in silicon under contact loading," Rev. Adv. Mater. Sci. 3 (2002) 1-36.

Novel Evaluation and Characterization

A PORTABLE MICROWAVE SCANNING TECHNIQUE FOR NONDESTRUCTIVE TESTING OF MULTILAYERED DIELECTRIC MATERIALS

Karl Schmidt
Evisive, Inc.
Baton Rouge, Louisiana, USA

Jack Little – Evisive, Inc.
Evisive, Inc.
Baton Rouge, Louisiana, USA

William A. Ellingson
ERC Company
Indianapolis, Indiana, USA

ABSTRACT

Determining when to replace current composite ceramic materials to preserve functionality is an important cost factor in their use. The question posed is, "How to estimate performance ability of the individual composite ceramic s in situ?" Identification of damaged material is important usefulness and replacing only damaged material is an important cost factor. Recently, a portable one-sided microwave scanning technique, that can be used in situ, has been demonstrated to detect damage on actual test specimens. Work has been conducted using as-produced and intentionally impact damaged specimens. Data from the microwave scanning test method have been compared directly to through-transmission x-ray examination data for the sample specimens. Each "crack" detected by through-transmission x-ray is also detected by the microwave scanning method. While the visual presentation of the microwave image is not identical to the x-ray image, it is equally unambiguous with respect to detection of features in the specimen. Based upon development work done to date, it should be possible to achieve microwave scanning rates which would permit inspection of an area of about 4 square feet (2 feet by 2 feet) in 5 minutes or less. It appears that the method is suitable for in-situ monitoring of materials.

INTRODUCTION

High-performance technical ceramics are widely used in field environments. Effectiveness of multilayered ceramics can be degraded by defects present from production and by operational damage resulting from handling or impact with objects in the environment. In normal use, multilayer ceramics are routinely exposed to the possibility of such damage.

A means to detect damage from handling and manufacturing defects is needed to determine the integrity of the product including identifying potentially degraded functionality. In order to have a timely and efficient condition assessment, it is desirable to monitor the material in-situ, with minimal access cost and without having to remove encapsulating materials. This would best be performed with access from one surface only, and without removal of mounted components. It would be desirable to have real time determination of service readiness by means of immediate nondestructive imaging and go/no-go condition assessment criteria. It is desirable to detect surface and internal cracks in the ceramic tiles, and laminar features in the surrounding material. The system and technique should be designed for broad use, requiring low cost, easy to use technology.

The objective of this activity was to demonstrate the feasibility of the application of a patented microwave scanning device as a portable NDE methodology to perform monitoring of the condition of multilayer ceramic material. The scanning process is referred to here as Evisive Scan TM. This method has been used successfully in preliminary studies on many different types of dielectric materials. It has been shown to be highly effective for quick, accurate detection of discontinuities in monolithic ceramics including samples of ceramic armor panels. The Evisive Scan image has been demonstrated to be useful in comparing test samples to a standard and in comparing images of a part before and following service induced degradation.

This method has been demonstrated on an armor panels provided by a variety of independent sources. The Evisive Scan method permits real time evaluation by inspection from one surface only, through non-contacting encapsulation.

THE EVISIVE SCAN METHODOLOGY

The Evisive Scan method uses a microwave transmitter to bathe the subject part in microwave energy, and uses multiple receivers to detect the integrated emitter and reflected signals. The method and equipment is covered by U.S. and international patents[1]. It utilizes microwaves as an interrogating beam to penetrate a dielectric material. The microwaves are reflected at areas of changing dielectric constant. This reflection and the interrogating beam combine to form an interference pattern. The microwave signal is measured by the receiver as voltage differences at locations over the surface of the material. The voltage pattern is displayed as an Evisive Scan image and indicates the presence of a potential defect, internal structure of interest, or change in material dielectric properties.

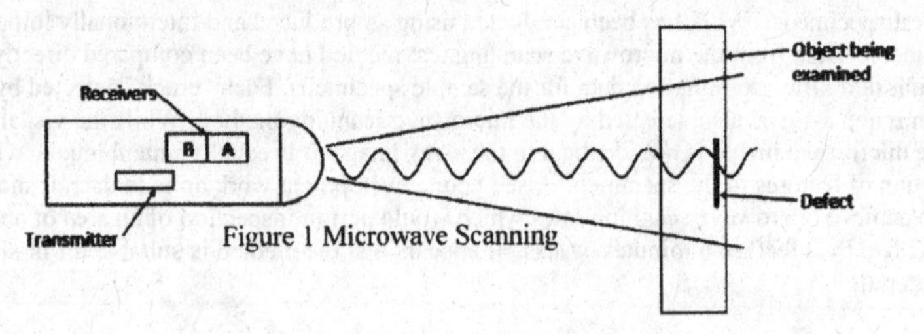

Figure 1 Microwave Scanning

Figure 2 Illustrates the combined signal which produces an interference pattern from a reflector at a depth in the material. The received microwave signal yields a point cloud which uniquely characterizes the volume of the inerrogated material. The difference between the emitter signal and reflection is a standing wave pattern in which the Z distance to a reflection within a wave length defines phase relationship of the reflected signal to the emitted signal and the amplitude of the difference signal. The difference is a variable DC voltage. This DC signal creates the image value for a given point over the part surface. This point cloud is the Evisive Scan data and yields the images directly. The Evisive Scan images presented here are generated directly from data files of the scan voltage.

Evisve Scan images are interference patterns. As such, the threoretical limts of feature detection are very small. In applications, the resolution of adjacent features in material is determined by the energy pattern of the reflected wave and the energy gathering profile of the specific antenna. as well as the gradient of the dielectric properties at the feature itself. The antenna geometry also defines the energy profile of the emitted wave and thus also impacts the reflected signals. For the apparatus configuration used in most of the reported tests. and the materials tested. this value is about 0.01 inches to 0.03 inches.

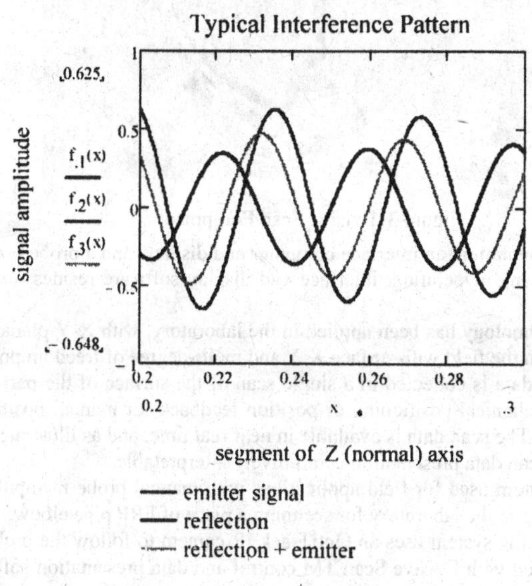

Figure 2 Interference Pattern DC

EQUIPMENT CONFIGURATION

The Evisive Scan TM equipment is shown in Figure 3. The equipment consists of a microwave

Figure 3 Evisive Scan Equipment

probe, processing instrumentation, an interface computer and display, and a probe positioning or tracking mechanism (not shown). Operating, interface and display software resides on the interface and display computer.

This scanning technology has been applied in the laboratory, with X-Y planar, X-Y cylindrical and r-θ positioning, and in the field with surface X-Y and multi-degree of freedom positioning devices.

The examination data is collected in a single scan of the surface of the part. The data rate is sufficiently high that mechanical positioning or position feedback for manual positioning is the only limitation in scan speed. The scan data is available in near real time, and as illustrated in the following images, the microwave scan data presentation is intuitively interpretable.

The Tracking System used for field applications with manual probe manipulation is shown in Figure 4, where it is set up in the laboratory for scanning a piece of FRP pipe elbow.

The position tracking system uses an OptiTrack IR camera to follow the probe position. Position data is fully integrated with Evisive Scan TM control and data presentation software. Some data was collected with a stationary x-y plotter table in the laboratory.

FLAW DETECTION

In the specimens evaluated to date, flaw detection has been determined by correlation of microwave scan data with visually observed defects in the scanned part, known morphology of artificially created defects and with nondestructive test data from alternative methods. X-ray images were made of artificially cracked sample pieces by Argonne National Laboratory and made available to support the work reported here.

Alternative nondestructive testing methods are ineffective in the presence of separation of en-

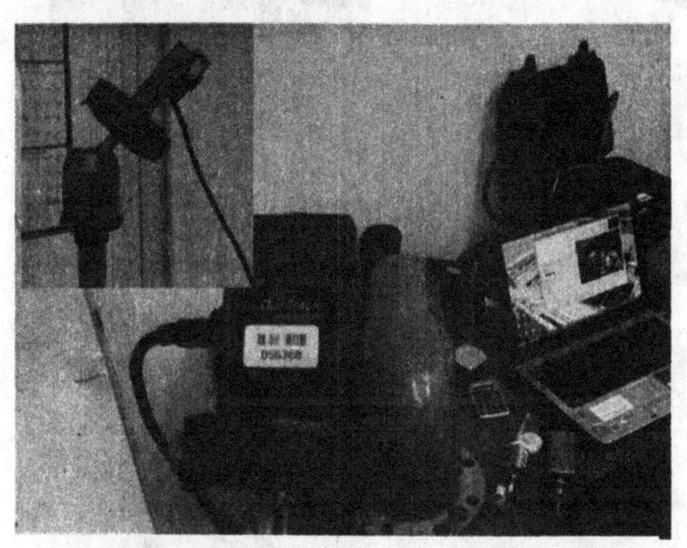

Figure 4 InfraRed Tracking System for Manual Probe Manipulation

capsulation from the inspected part surface[2]. Alternative radiographic nondestructive testing methods are limited by requiring access to both sides of the specimen, by face contact (two dimensional flaw morphology) in flaws and by flaw orientation relative to the part surface plane. Alternative thermographic inspection methods are limited by encapsulation materials and face contact (two dimensional flaw morphology) in flaws.

The Evisive Scan technology has been successfully applied to multilayer ceramic specimens provided by third parties.

The following figures include images of samples provided to Evisive by third parties. The material was supported here by a ¼ inch steel plate. Access was accommodated only from one surface. The specimen was examined at Evisive before and after it was cracked in the laboratory by providing party.

CRACKED TILE IMAGES

The images in Figure 5 clearly show the crack pattern and the reduced ceramic section in the cracked tile. From the top: a photo image of the surface of the composite encapsulation from which the Evisive Scan examination was made, the Evisive Scan images of the cracked tile, an x-ray of the tile and an operator interpretation of the Evisive Scan image, illustrating the identified cracks. The cracks show as pairs of lines characteristic of the reflection interference pattern from the well-defined

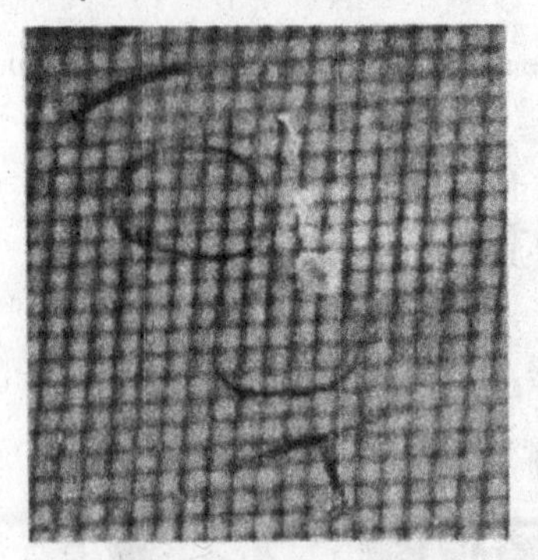

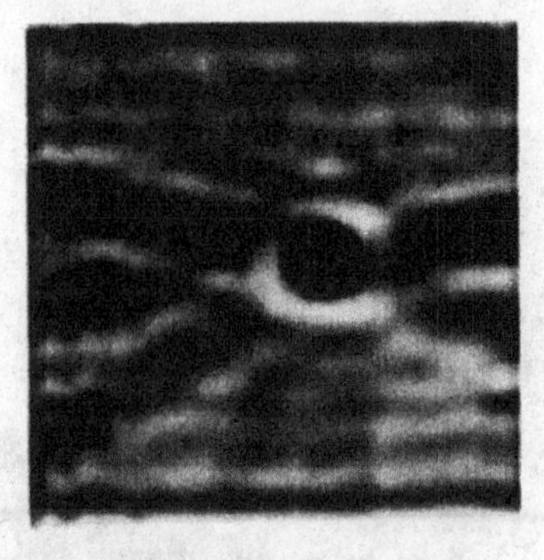

Figure 5(a) Photo Image of Panel Face and Microwave Scan of Cracked Tile

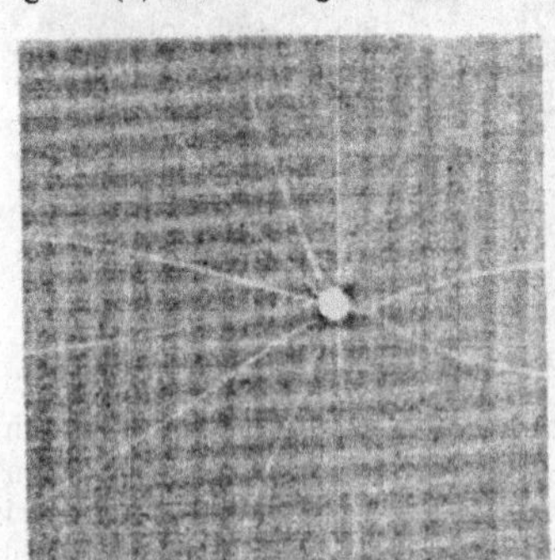

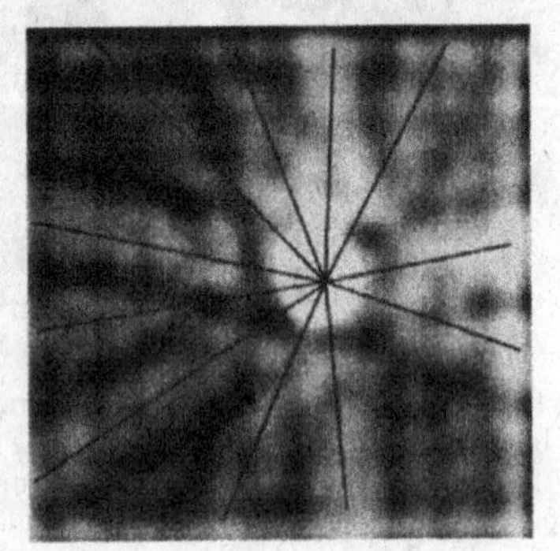

Figure 5(b) X-Ray Image and Microwave Scan Image with Interpretation of Cracked Tile

discontinuity. Precise measurements can be made from the image in real time by the operator. The measurements and parameters of features analyzed by the operator are saved with the Evisive Scan TM data file. The images above are about four inches square. This image compares precisely with the x-ray image of the cracked tile. The x-ray image was created by examining the part with through transmission, an x-ray source requiring access to both surfaces of the part. The artifically added red lines in the bottom figure indicate the operator interpretation of crack locations in the Evisive Scan TM image. The lines were added to show the orientation and placement of cracks identified in the Evisive

Scan image and are clearly coincident with the defects detected in the x-ray image of the part. Actual interpretation of the Evisive Scan TM is documented in the image data file and is explicit for each artifact analyzed by the operator. The cracks are clearly visible by the paired interference pattern reflection of the permeability change at the crack surface. The rectilinear pattern in the Evisive Scan image is the reflected edge of the tile. The x-ray is shown in "negative" format. The cracks appear as thin white lines.

This experiment demonstrates that the Evisive Scan TM technique can detect cracking ceramic tiles, with access from one side of a multilayered part; and that the scan image is substantially consistent with x-ray imaging of the part using transmission through the part (access from both sides).

The images in Figure 6, to the right, present a similar experiment conducted with a multilayered ceramic of different construction. The surface of the overwrap, seen in the upper image, is marked by the impact blow in the upper right quadrant. The pattern of fracture lines and the irregular penetration of the tile are clearly visible in the Evisive Scan image (middle). The image is consistent with the x-ray image made of the same part (bottom) using access from both surfaces. The x-ray image is presented in "negative" format.

The tile in the upper right position was cracked using an impact blow in the laboratory which provided the material. The prominent vertical and horizontal pattern in the Evisive Scan image in the Evisive Scan image is the interference pattern image of the rectilinear edges of the tiles. The lower left tile in Figure 6 also exhibits a voltage difference (gray scale difference) suggesting a feature in the encapsulating material, such as an absence of material, inclusion of foreign material or a disbond. Such features are explored in the next section.

The varying density pattern of the tile area in the x-ray image is the sum of the weave patterns of the front and back encapsulation fabrics. This is also visible in the x-ray image in Figure 5. Also in that image are several areas of reduced material density (lighter irregular shapes in the tiles above and below the cracked tile). These indicate a relative absence of material in the locations.

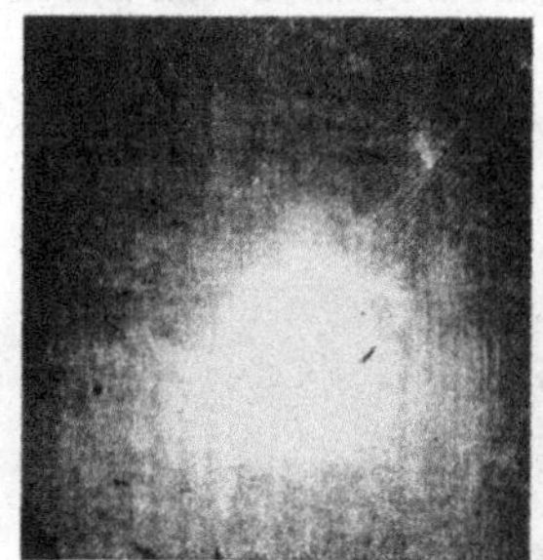

Figure 6(a) Photo Image of Panel Face

Figure 6(b) Microwave Image of Panel

Figure 6(c) X-ray Image of Panel

Calibrated position measurements can be made from the microwave image, enabling analysis of observed artifacts. The Evisive Scan image was made using the scanning table, with access from one surface only. The part was backed with ¼ inch steel. Alternative positioning means such as the IR Tracking system illustrated in Figure 4 (replacing the scanning table) have been demonstrated for field use. The images above are about 8 inches square.

DISBONDS AND OTHER FEATURES

The following figures illustrate detection and imaging of features in the ceramic composite material other than cracks in the tiles. Figure 7 is a false color presentation of Channel A data from a segment of the large scan from which he image shown in Figure 5 was extracted. The color spectrum associated with data voltages is set by the operator in the display. As described in Evisive Scan Methodology. and illustrated in Figure 2, Channel A and Channel B images represent the reflected

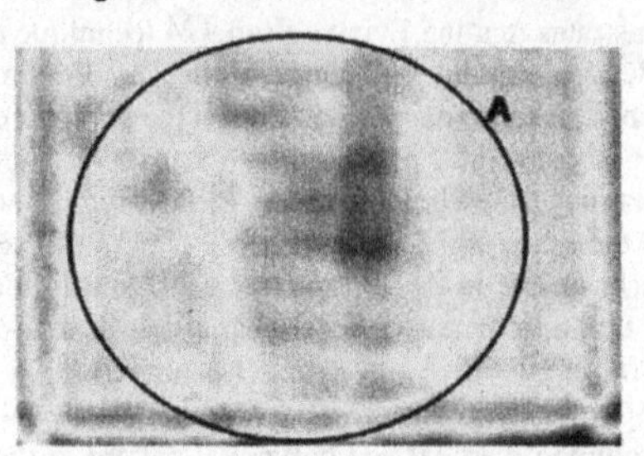

Figure 7 Specimen Scan Channel A

signal interference pattern with a phase difference of ¼ wave length at a specific depth in the material which is imaged. We shall see in Figures 8, 9 and 10, that the phase differential yields significant information about the depth of features in the material, and that multiple features at different depths can be resolved (stacked features). Because the microwave signal is reflected at every change in dielectric property, laminar features without volume, such as disbonds; may be readily imaged. The image identified as A in Figure 7 is prevalent in Channel A and absent in Channel B, indicating that the associated feature has dielectric properties which are very much related to depth, such as a disbond in the layered material.

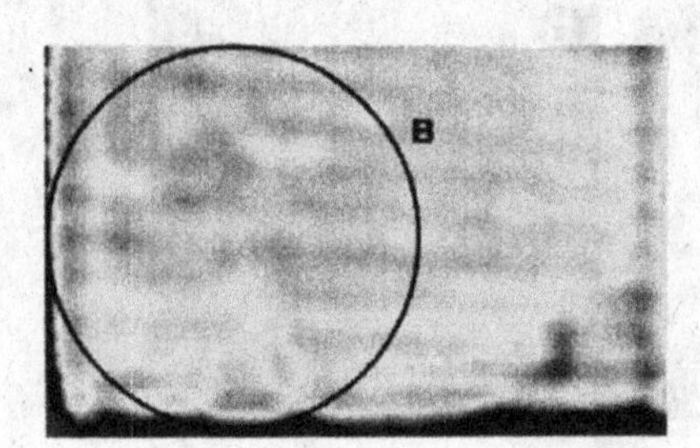

Figure 8 Specimen Scan Channel B

Figure 8 is Chanel B of the same scan of the specimen. This Channel is separated in phase from Channel A (figure 7) by ¼ wavelength. Thus, features that are located at a different depth in the material are more apparent here. This is illustrated by the presence of the irregular features labeled B, which occur in the same X-Y location as the feature labeled A (in Figure 7). The features are at different depths in the material, presenting themselves more vividly in the different channels. In comparison with the x-ray image (shown in Figure 10) it is apparent that the material is of relatively uniform density in these regions. The images in Figures 7 through 11 are about 2 inches by 4 inches.

The regular rectilinear pattern in Figures 7, 8 and 9 is the lay of fiber in the encapsulating material. The near and far surface fiber matrices can be differentiated by the relative response in the separate Channels.

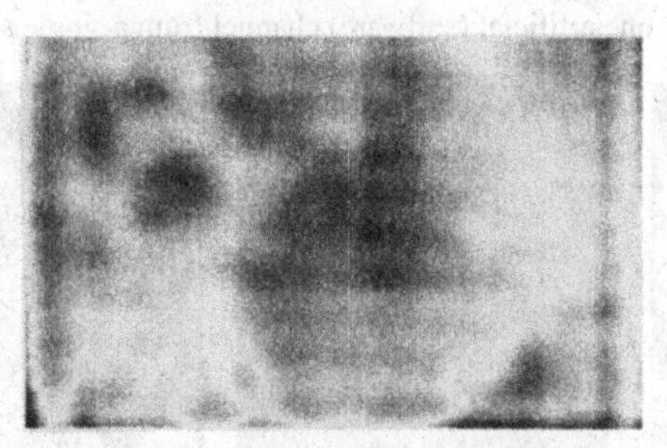

Figure 9 Specimen Scan Channel C

The fiber is pattern is apparent in Channel A, B and composite Channel C images. This illustrates the ability of the microwave interference measurement to resolve features substantially smaller than the wavelength (here about 12.7 mm (0.5 inches)). The spacing of the fiber ribbons is about 3.2 mm (0.125 inches). and the fiber ribbons themselves are about 3.2 mm (0.125 inches) wide.

The x-ray image "negative" of the specimen is included here as Figure 10 for comparison with the Evisive Scan images. Note the uniformity of the specimen density and the faintly visible fiber pattern. The surface image of the part is included as Figure 11.

Figure 10 Specimen X-Ray Image

Figure 11 Specimen Face Image

STACKED FEATURES

In the following figure, several features are visible at different depths in the material in same image and the different depths can be inferred from the data recorded in the in the different phases of the two hardware channels and the one artificial (software) channel from a single scan of the part.

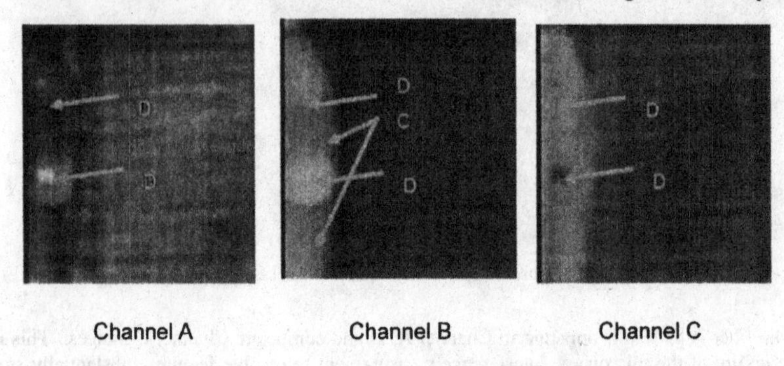

Figure 12 Stacked Features in Channels A, B, and C

The elongated feature (labeled C) on the left edge of each image is a separation of the matrix surrounding the ceramic tile. It is most readily noted in Channel B as white (more positive voltage than null in conventional display settings). The round features (labeled D) are at a different depth as indicated by the substantially voltage difference in A and B channels (in C they are dark, negative relative to null). The round features are likely to be small voids in the matrix material which have been compressed into a disbond with little or no volume. The images in Figure 12 are about two inches square.

CONCLUSION

The Evisive Scan microwave interference scanning technique has been demonstrated to be suitable for examination of multilayer ceramic composite material. Examination requires access from one side only and is effective in applications with metal backing.

Experiments to date have demonstrated ability to detect and image: cracks in ceramic tiles, inclusions, foreign materials and laminar features in the composite matrix. Experiments have demonstrated imaging and characterization of features with dimensions well below the wave length of the transmitted microwave signal.

Further laboratory testing and evaluation of known defects and features is required to establish characterization and interpretation protocols and qualify the technique for nondestructive testing applications.

REFERENCES

[1] UNITED STATES AND INTERNATIONAL PATENTS
"APPARATUS AND METHOD FOR NONDESTRUCTIVE TESTING OF DIELECTRIC MATERIALS", U.S. Patent 6,359,446. MAR. 19.2002
"INTERFEROMETRIC LOCALIZATION OF IRREGULARITIES U.S. Patent 6.653,847, Nov. 25, 2003
"HIGH-RESOLUTION, NONDESTRUCTIVE IMAGING OF DIELECTRIC MATERIALS", PCT/US2005/026974, International Filing Date 1 August, 2005
"NONDESTRUCTIVE TESTING OF DIELECTRIC MATERIALS", Canadian Patent 2,304,782, Mar. 27, 2007
"NONDESTRUCTIVE TESTING OF DIELECTRIC MATERIALS", Australian Patent 746997, New Zealand Patent 503733
PCT/US2005/026974, International Filing Date 1 August, 2005

[2] "PROGESS IN CERAMIC ARMOR", Transactions of the Am. Cer. Soc, Vol. 134, 2002

[3] "CERAMIC ARMOR AND ARMOR SYSTEMS II", edited by Eugene Medvedovski, Ceramic Transactions of the Am. Cer. Soc Vol. 178, 2005

BALLISTIC DAMAGE ASSESSMENT OF A THIN COMPOUND CURVED B_4C CERAMIC PLATE USING XCT

J.M. Wells[1] and N.L. Rupert[2]

[1]JMW Associates, 102 Pine Hill Blvd, Mashpee, MA 02649; 774-836-0904;
jmwconsultant@comcast.net

[2]KLNG Enterprises, Inc., 122 Bradish Road, Kittanning, PA 16201-4306; nevinlr@comcast.net

ABSTRACT

Not all ceramic armor plates are flat. The current study looks at characterizing ballistic impact damage in a thin boron carbide, B_4C, body armor torso plate with a compound curvature impacted by lead core rifle ball ammunition. The non-invasive characterization and visualization of this impact damage was conducted post-mortem using previously established x-ray computed tomography, XCT, diagnostic techniques [1-11].

Despite being impacted from three separate hits, the ceramic plate remained essentially intact. Our initial and in-situ damage observations include: large cratering on the impacted B_4C front surface, three complete penetrations of B_4C plate but apparently not through the composite backing plate, substantial bulging on the rear of the composite backing plate at each hit location, extensive mesoscale cracking and ceramic fragmentation in the B_4C plate, and multiple small high density projectile fragments embedded in the composite backing plate. These qualitative and quantitative ballistic impact damage initial results are reported and discussed along with 2D and 3D virtual images. The results are considered significant in permitting the cognitive visualization of such in-situ impact damage and thus aid in improving our understanding of both the physical damage and penetration manifestations in such a target.

INTRODUCTION

The non-invasive x-ray computed tomography, XCT, diagnostics of impact damage in terminal ballistic laboratory target materials of relatively simplistic shapes has been explored and reported by the authors and their collaborators for the past decade [1-11]. In the present work, the authors have applied the non-invasive XCT diagnostic modality to the inspection of ballistic impact damage in a commercially produced compound curved body armor torso plate assembly consisting of a B_4C ceramic plate adhered to an organic composite backing plate. This target had been impacted by three successive shots of lead core rifle ball ammunition prior to our examination. The purpose of this work was not to qualify the body armor target materials or configuration, but rather, to demonstrate the feasibility of the XCT inspection modality to provide a more comprehensive, yet still non-invasive, characterization of resulting internal impact damage in a realistic body armor component.

BACKGROUND

Currently, the quality assurance inspection techniques generally applied during the manufacturing stage of body armor ceramic components are visual, metrological, and projection x-ray (i.e. film or digital) modalities. In-service or post-battle damage inspection results are not well documented in the open literature. In any case, internal inspections, before or after impact, are normally conducted either with destructive and/or non-destructive 2-dimensional approaches.

Similarly, 2-D examination modalities are most frequently utilized in the laboratory research and development stages of various armor materials and their architectural designs.

Ultimately, the bottom line qualification focus for body armor components is on the results of prescribed ballistic test requirements for penetration. Penetration resistance is admittedly a most necessary requirement and the best accepted proof test for armor applications. However, penetration data by itself does not provide an adequate characterization of the constituent armor materials damage behavior, nor does it provide explicit guidance toward necessary and quite desirable weight reductions and material improvements. A critical material constituent in most Type III or Type IV rigid military body armors is a compound curved solid ceramic plate which is quite brittle and experiences considerable physical damage. Because of its low density and its attractive ballistic performance, the most common ceramic used for this application is B_4C.

TARGET DESCRIPTION

The subject rigid body armor component examined was an archival B_4C ceramic striking plate adhesively bonded to a fiber reinforced organic composite backing plate. No specific item specifications or identification markings are presented, nor is information made available here on the plate origin or the specific ballistic test conditions to which it was subjected. It is, however, confirmed that the projectile type used was NIJ Level III [12] lead core rifle ball ammunition and that three separate shots were made according to standard ballistic testing procedures.

XCT DAMAGE DIAGNOSTIC APPROACH

The subject body armor plate was scanned in the horizontal position such that the axial scanning direction was perpendicular to the armor center line thickness direction. A total of 45each 2-D axial scans constitute the original XCT scan file. The scan slices are of 16 bit unsigned data file format. The voxel resolution is calculated as 0.160 mm^3.

The requisite post-scanning reconstruction and various image processing, analyses, and visualizations were conducted using Volume Graphics StudioMax (VGSM) v.1.2.1 voxel analysis software [13]. Initially, the original volumetric digital scan data is imported into the VGSM software and then quickly rendered to reconstruct a virtual 3-D solid object from which all further image processing results and additional 3-D visualizations are derived. This 3-D solid object can then be rotated to any desired observation position, adjusted in magnification, and segmented into various grey level (density) regimes. The orthogonal 2-D cross-sectional images are also available to observe consecutive virtual sectioning in the axial (X-Y plane), frontal (X-Z plane), and sagittal (Y-Z plane) directions. Other 2-D virtual sectioning planes can also be produced at arbitrary orientations as desired. All rendered images are saved in the tiff image format.

RESULTS

A virtual 3-D solid object reconstruction of the subject armor plate assembly is shown in fig.1 with the locations of the three hits designated as top, mid, and bottom. Also shown in this figure are the overall length and width measurements made in-situ within the virtual metrology capabilities of the VGSM software. The length of this plate measured as ~318mm and the width as ~239mm in-situ on the virtual 3-D solid object reconstruction of B_4C ceramic striking plate. The concentric oval rings observed on the ceramic surface are actually scanning artifacts reflecting the original incremental axial scan levels conducted sequentially through the object thickness. This is actually a somewhat serendipitous case as these artifacts assist in visualizing the complex curvature of this plate in a fashion similar to the topological features on surface map.

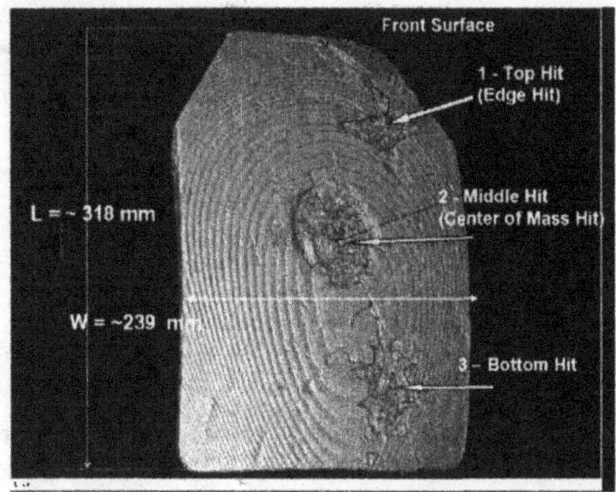

Figure 1. 3-D Solid Object rendering of the B_4C/Composite Armor Plate

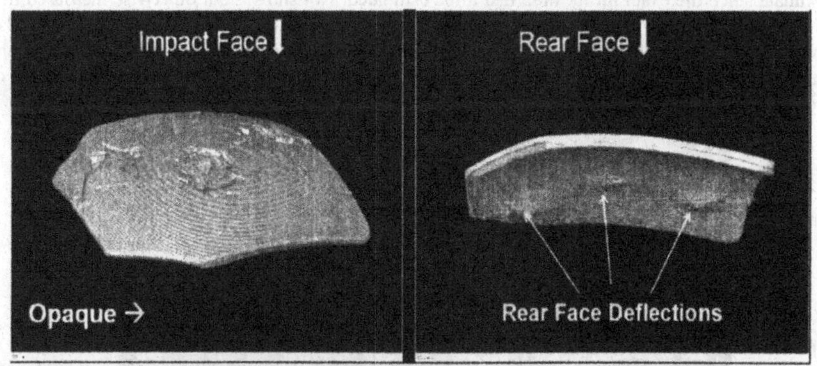

Figure 2. Oblique view of the B_4C impact face (left) and the composite rear face (right).

Two additional 3-D views of this object are shown in an oblique orientation in figure 2. The absence of some ceramic material caused by ceramic fragmentation and cratering on the front impact surface at the three impact locations is observed in both figures 1 and 2. Also seen in figure 2, are three distinct bulges on the rear face of the organic composite backing plate, coincident with the three ballistic impact locations. A 2-D cross sectional axial scan shown in figure 3 confirms the presence of two separate adjacent contoured plate materials with distinctly different normal densities as indicated by their respective grey levels.

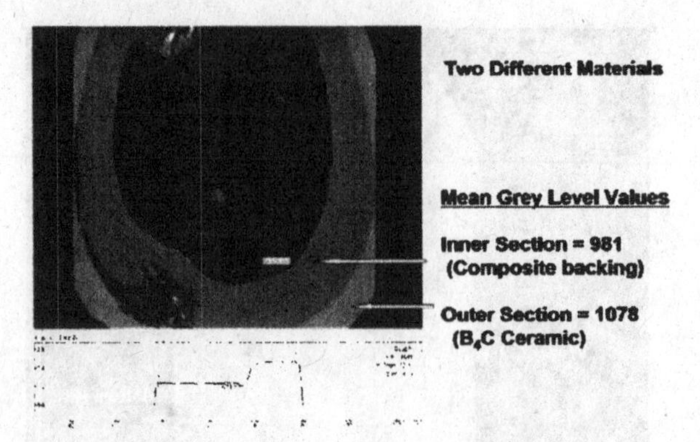

Figure 3. Axial 2-D cross section scan #22 confirms the presence of two different plate materials by their respective grey levels.

The in-plane spacing of the individual hits from the approximate center-of-mass (Mid) hit is shown in figure 4 as revealed by in-situ linear metrology on axial plane scan A-25. These spacing distances are approximate since the exact hit location had to be estimated. Not shown, but otherwise measured in a transverse frontal slice was the in-plane distance from the top to the bottom hit of ~190mm. Also observed in the images in figure 4 is the obvious and substantial interface separation between the ceramic and composite backing plates from the top hit all the way to the bottom hit on the left side of the armor plate.

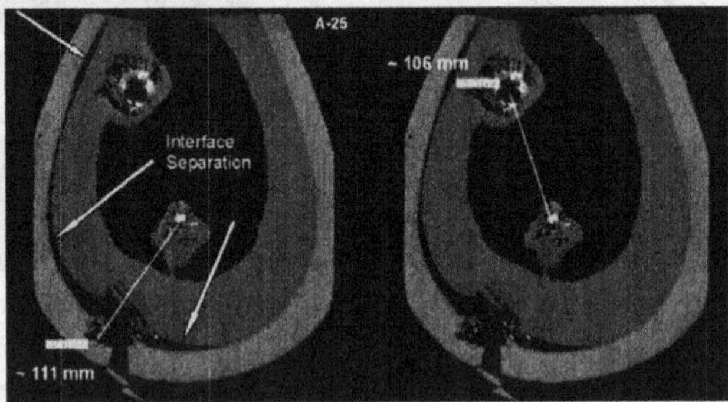

Figure 4. Axial slice #25 showing metrology of hit spacing – Top to Mid ~111mm (left) and Mid to Bot ~106 mm (right). Interface separation and debonding are seen at arrows.

Multiple embedded bullet fragments are found at each hit location as shown in the semi-transparent 3-D oblique image at the top of figure 5. A further perspective of these high density fragments can be observed in the selected 2-D axial scan image planes progressing from the impact toward the rear face at the bottom section of figure 5. In these axial figures it becomes apparent that all of the fragments observed here at each impact location reside not in the B_4C ceramic plate, but rather within the composite backing plate.

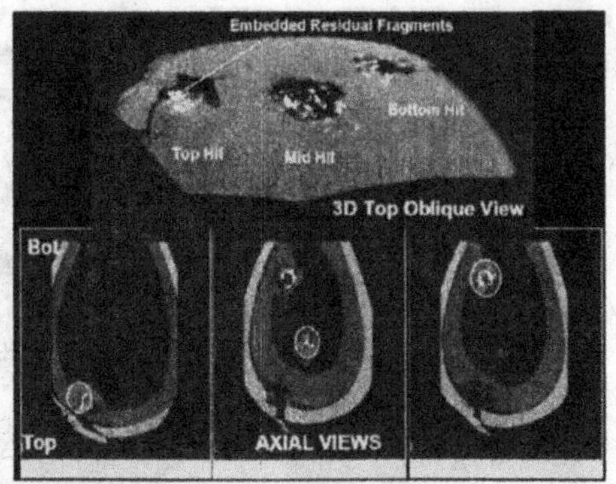

Figure 5. Semi-transparent 3-D oblique view (upper) and opaque 2-D axial views (lower) showing multiple bullet fragments circled in the backing plate at the 3 hit locations.

In order to adequately inspect the rear surface of the B_4C ceramic plate, it is necessary to virtually remove the opaque composite backing plate. This was accomplished by means of a density segmentation where all grey level values below that of the ceramic plate are discarded and the 3-D solid object is rendered for the higher grey level values only. The result of this segmentation operation makes the backing plate virtually transparent and thus the concave ceramic rear surface is directly visualized.

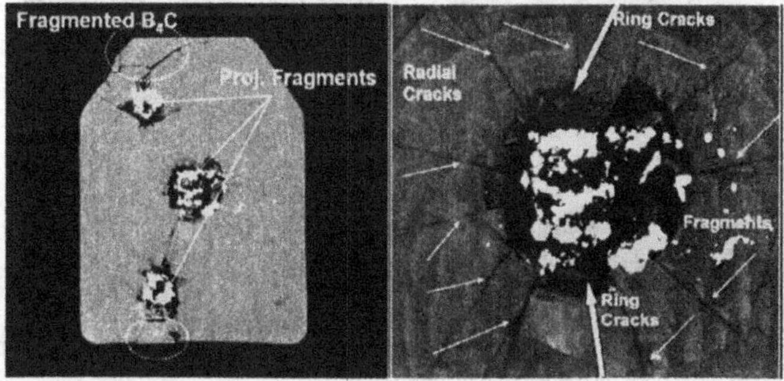

Figure 6. Segmented view of rear face of B_4C plate without the composite backing (left) and an enlarged view of mid hit location (right) showing cracking in the B_4C plate and multiple projectile fragments visible in the volume occupied by the absent backing plate.

Without the opacity of the backing plate obstructing the view of the ceramic rear surface, it is now possible to directly observe both sides of the B_4C ceramic plate unobstructed, as shown in figures 6 and 7. Substantial cracking damage and intact ceramic fragmentation exists on both surfaces particularly in the three near impact locations. A large irregularly shaped through hole exists at each hit location, with a considerable volume of ceramic material displaced and missing. An enlarged view of the radial and ring cracking observed at the mid hit location on the ceramic rear surface is shown in figure 6, along with the bullet fragments in virtual suspension revealing a noticeable size and spatial distribution.

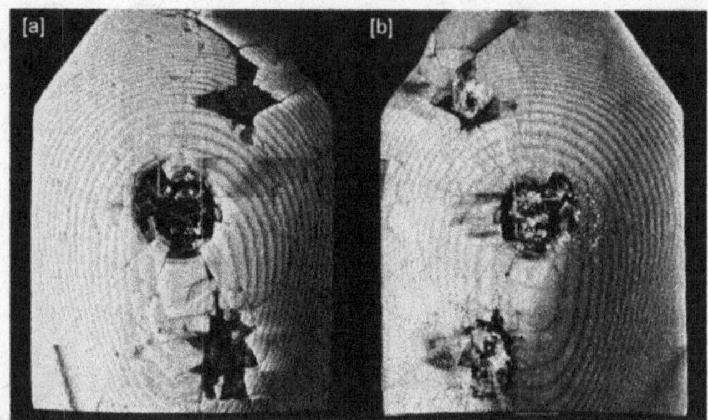

Figure 7. Front side (a) and rear side (b) of B_4C plate showing through holes in the ceramic and residual bullet fragments shadowed at all three hit locations with backing plate removed by density segmentation (opaque view).

Figure 8 shows a closer view of the interconnected radial cracks between the top and bottom hit sites with the mid hit site. Also shown on the right side of this figure is an

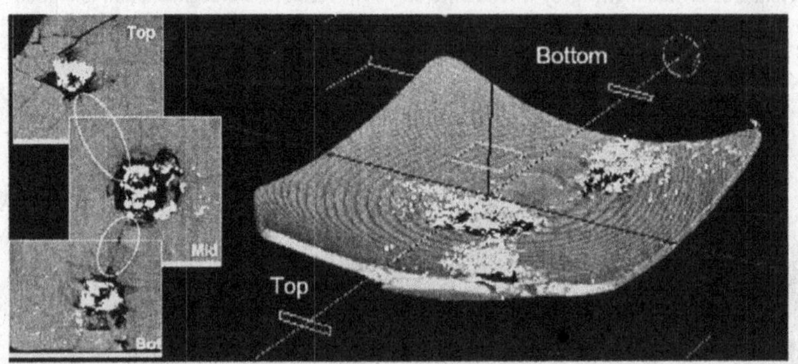

Figure 8. Interconnecting cracking between hits (left) and 3-D view of fragments with composite backing virtually removed on the concave rear face (right).

alternative perspective for the 3-D visualization of the many small distributed residual bullet fragments in virtually suspension above the concave B_4C ceramic rear surface. It is planned to conduct further analyses, both qualitative and quantitative on the several damage manifestations presented above.

SUMMARY

The application of established XCT damage diagnostic techniques were applied to the initial noninvasive inspection of a ballistic impacted complex curved B_4C torso body armor plate. Full perforation of the B_4C plate with considerable ceramic damage was observed at all three hit locations. Substantial ceramic fragmentation occurred adjacent to the top and bottom hit locations but remained attached. Significant areas of unattached and displaced ceramic fragmentation leaving a large through-hole occurred at each hit location. Separation distances between the three hit locations were measured in-plane with virtual in-situ metrology. Direct examination of the "hidden" concave rear surface of the ceramic plate was accomplished by the grey level segmentation and virtual removal of the composite backing. Both radial and ring cracks were observed on the rear surface of the ceramic plate at the impact locations. Interconnecting radial cracks are present between the middle hit and the top and bottom designated hit locations respectively. Separation of the adhesive interface between the ceramic and the backing plates occurred on the left side of the object where the three hits were concentrated. Once such interface separation occurs, it is reasonable to suspect that it could reduce subsequent ballistic impact performance of the system. The authors suggest that this interface separation phenomenon needs to be measured and further characterized as much as the physical ceramic damage. Prominent plastically deformed bulges were observed on the rear side of the composite backing plate at each hit location. Multiple small bullets fragments are present in the backing plate at each hit location and were best observed by segmentation and the virtual removal of the backing plate.

While additional examination and in-situ metrology actions are still planned for this body armor plate, the initial results shown here are considered most encouraging to demonstrate the functional capability of noninvasive XCT diagnostics for the post-mortem terminal ballistic damage analysis of actual body armor.

REFERENCES

[1] J.M. Wells. On the Role of Impact Damage in Armor Ceramic Performance. Proc. of 30th Int. Conf. on Advanced Ceramics & Composites-Advances in Ceramic Armor, 2006.

[2] J.M. Wells. Progress in the Nondestructive Analysis of Impact Damage in TiB_2 Armor Ceramics. Proc. of 30th International Conf. on Advanced Ceramics & Composites-Advances in Ceramic Armor. 2006.

[3] J. M. Wells. Progress on the NDE Characterization of Impact Damage in Armor Materials. Proc. of 22nd Int. Ballistics Symp.. ADPA. v2. pp. 793-800. 2005.

[4] H.T. Miller, W.H. Green, N. L. Rupert, and J.M. Wells, Quantitative Evaluation of Damage and Residual Penetrator Material in Impacted TiB_2 Targets Using X-Ray Computed Tomography. 21st Int. Symp. on Ballistics, Adelaide. Au. ADPA, v1, pp. 153-159, 2004.

[5] J. M. Wells. N. L. Rupert. and W. H. Green, Progress in the 3-D Visualization of Interior Ballistic Damage in Armor Ceramics. Ceramic Armor Materials by Design, Ed. J.W. McCauley et al.. Ceramic Transactions, v. 134, ACERS, pp. 441-448. 2002.

[6] J.M. Wells. On Incorporating XCT into Predictive Ballistic Impact Damage Modeling. Proc. of 22nd Int. Ballistics Symp.. ADPA. v2. pp. 1223-1230. 2005.

[7] J.M. Wells, On Continuing the Evolution of XCT Engineering Capabilities for Impact Damage Diagnostics., Proc. 31st Intn'l Conf. on Advanced Ceramics & Composites. ACERS, 2007.

[8] J.M. Wells, N.L. Rupert, W.J. Bruchey, and D.A. Shockey, XCT Diagnostic Evaluation of Ballistic Impact Damage in Confined Ceramic Targets. 23rd Intn'l Symp. on Ballistics, Tarragona, Spain. ADPA v2. pp. 965-972. 2007.

[9] N.L. Rupert. J.M. Wells. W. Bruchey. and J.R. Wheeler, The Evolution and Application of Asymmetrical Image Filters for Quantitative XCT Analysis. 23rd Intn'l Symp. on Ballistics, Tarragona, Spain, ADPA v2. pp. 945-952, 2007

[10] J.M. Wells. On the Linkage of Impact Damage to Modeling of Ballistic Performance. Computational Ballistics III, ed. C.A. Brebbia and A.A. Motta, WIT Press. PP 89-98, 2007.

[11] J.M. Wells and R.M. Brannon. Advances In X-Ray Computed Tomography Diagnostics of Ballistic Impact Damage, Metallurgical and Materials Transactions A. v. 38A, pp 2944-2949, 2007.

[12] National Institute of Justice, Ballistic Resistance of Personal Body Armor. NIJ Standard-0101.04. Washington, DC: U.S Department of Justice. National Institute of Justice, June 2001, NCJ 211680.

[13] http://www.volumegraphics.com/

EVALUATION OF BALLISTICALLY-INDUCED DAMAGE IN CERAMIC TARGETS BY X-RAY COMPUTED TOMOGRAPHY

William H. Green[1], Herbert T. Miller[1*], Jerry C. LaSalvia[1], Datta P. Dandekar[2], and Daniel Casem[2]
U.S. Army Research Laboratory
Weapons and Materials Research Directorate
AMSRD-ARL-WM-MD[1]/TD[2]
Aberdeen Proving Ground, MD, USA

ABSTRACT

X-ray computed tomography (XCT) is an important non-destructive evaluation technique for revealing the spatial distribution of ballistically-induced damage in ceramics. The level of detection and resolution of damage depends on the size of the specimen and the parameters of the XCT approach (e.g., focal spot size, magnification, etc.). Previous and ongoing work in assessment of ballistically induced damage in ceramic targets includes the study of both directly impacted and plane shock wave damaged specimens. Shock induced damage in both AD995 alumina and SiC-N ceramic targets have been scanned and extensively evaluated using XCT 2-D and 3-D analysis. The damage in the AD995 alumina and SiC-N targets was induced by 6 GPa and 4 GPa pressures, respectively. The purpose of using XCT evaluation in these targets is to better characterize and understand all of the detectable damage and possibly correlate damage features and types with the physical processes of damage initiation and growth. XCT scans of shock induced damage in ceramic targets will be shown and discussed. This will include virtual 3-D solid visualizations and some quantitative analysis of damage levels (i.e., decimal fraction of 100 percent) and particular damage features.

INTRODUCTION

The non-destructive x-ray computed tomography (XCT) technique is a widely applicable and powerful inspection modality for evaluation and analysis of geometrical and physical characteristics of materials. It also presently appears that the non-invasive diagnostic approach with XCT provides the only sufficiently effective modality for high resolution induced shock and damage interrogation, spatial characterization, quantification, visualization and three-dimensional (3-D) analysis. The XCT shock/impact damage diagnostic approach has already been successfully demonstrated on opaque armor ceramics, TiC, TiB_2, SiC, and Al_2O_3, as well as Ti-6Al-4V metallic armor materials [1-17]. In this paper the qualitative and quantitative evaluation of shock damage in Al_2O_3 and SiC disk specimens is discussed. The full capabilities of the XCT diagnostic approach have not yet been reached and the beneficial utilization of this new volumetric impact knowledge has yet to be extensively applied and exploited. Further, this new volumetric damage knowledge has yet to be consistently utilized, if at all, in ballistic damage models for comparison to the models and subsequent analysis [18-19].

SHOCK RECOVERED Al_2O_3 (AD995) AND SiC (SiC-N)

Design of the shock recovery experiments ensured that the ceramic specimens were subjected to a single shock wave compressive stress of a predetermined magnitude followed by release of the compressive stress before the effects of lateral stress release distorted the compressive and release histories especially in the central regions of the specimens. Further, the design ensured that no tension developed in the specimens due to wave interactions. Thus the observed features in the recovered specimens could be unambiguously associated with the shock induced planer shock compression and release [3, 20]. The diameter and thickness of the Al_2O_3 specimen was 17.0 mm and 6.0 mm, respectively, and the diameter and thickness of the SiC specimen was 38.1 mm and 6.0 mm, respectively.

X-RAY COMPUTED TOMOGRAPHY

X-ray computed tomography (XCT) is broadly applicable to any material or test object through which a beam of penetrating radiation may be passed and detected, including metals, plastics, ceramics, metallic/nonmetallic composite material, and assemblies. The principal advantage of XCT is that it provides densitometric (that is, radiological density and geometry) images of thin cross sections through an object in a non-invasive manner. Because of the absence of structural superimposition, images are much easier to interpret than conventional radiological images. The user can quickly learn to read XCT data because images correspond more closely to the way the human mind visualizes 3-D structures than 2-D projection radiology (that is, film radiography, real-time radiography, and digital radiography). Further, because XCT images are digital, the images may be enhanced, analyzed, compressed, archived, input as data to performance calculations, compared with digital data from other nondestructive evaluation modalities, or transmitted to other locations for remote viewing, or a combination thereof.

XCT SCANNING PROCEDURES

The semi-circular AD995 specimen, which was produced by sectioning the damaged disk, was carefully removed from its surrounding cup and scanned using the 225 keV microfocus x-ray tube and image intensifier (II)/charged coupled device (CCD) camera detector combination set up. The slice thickness and increment were .044 mm and .040 mm, respectively, resulting in overlapping scans. Each slice was reconstructed to a 1024 by 1024 image matrix. The tube energy and current used were 200 keV and .037 mA, respectively, and the focal spot was 5-µm. The circular SiC-N specimen was scanned in its surrounding cup using the 420 keV x-ray tube and linear detector array (LDA) set up. The slice thickness and increment were .500 mm and .500 mm. Each slice was reconstructed to a 512 by 512 image matrix. The tube energy and current used were 400 keV and 2.0 mA, respectively, and the focal spot was .80 mm. The SiC-N specimen was scanned using a significantly higher energy and thus a significantly larger spot size in order to penetrate its highly (x-ray) attenuating surrounding cup.

EVALUATION OF Al_2O_3 SPECIMEN

XCT Results

Figures 1.a to 1.i show a series of scans (slices) of the specimen starting at near the impact side and ending at near the rear side. Figure 1.a shows the crack damage about .20 mm from the impact side of the specimen. The notch in the straight edge is due to missing ceramic material that came loose sometime during handling. The cracking is mostly radial with one main circumferential lateral type crack towards the curved (semicircle) outer edge. This is the nature of the crack damage up to about 1.60 mm from the impact side. The damage is more severe about 2.20 mm from the impact side with multiple lateral cracks that are wider and jagged as well as being farther away from the semicircular outer edge, as shown in Figure 1.b. There is also less radial cracking at this location with some of it starting and ending at the lateral cracks and edge of the specimen, respectively. At about 2.50 mm from the impact side the most severe lateral cracking is about halfway between the semicircular and straight (sectioned) edges of the specimen, as shown in Figure 1.c. At about 2.70 mm from the impact side the circumferential lateral cracking is mainly in the center area of the specimen adjacent to the straight edge, as shown in Figure 1.d. The damage at this location is a concentrated distribution of relatively very wide and short cracks and low density, meaning less than undamaged nominal density, or void like regions. Secondly, longer radial cracks than in Figures 1.b and 1.c also extend from the concentrated damage to the semicircular edge. The cracking damage is less severe again farther away from the impact side, such that at about 3.10 mm distance there is very little lateral cracking with some radial cracking present, as shown in Figure 1.e. Figure 1.f shows the cracking damage at about 3.60

mm from the impact side. The lateral and radial cracking damage is more severe again at about 4.30 mm from the impact side, as shown in Figure 1.g. Similar to between 2.50 mm and 2.70 mm the cracking damage becomes more severe farther away from the impact side until at about 4.60 mm distance the majority of the damage is in the center area in the form of wide and short cracks and low density or void like regions again, as shown in Figure 1.h. Also similar is the decrease in the severity of the cracking damage past 4.60 mm. The second concentrated distribution of centered damage is no longer present with only a relatively low level of lateral and radial cracking present at about 5.10 mm from the impact side, as shown in Figure 1.i. This is the nature of the cracking through the rest of the thickness to the rear side. The lateral cracking again has a more circumferential nature in this region. The notch in the corner of the specimen in some of the images is due to missing ceramic material that came loose sometime during handling.

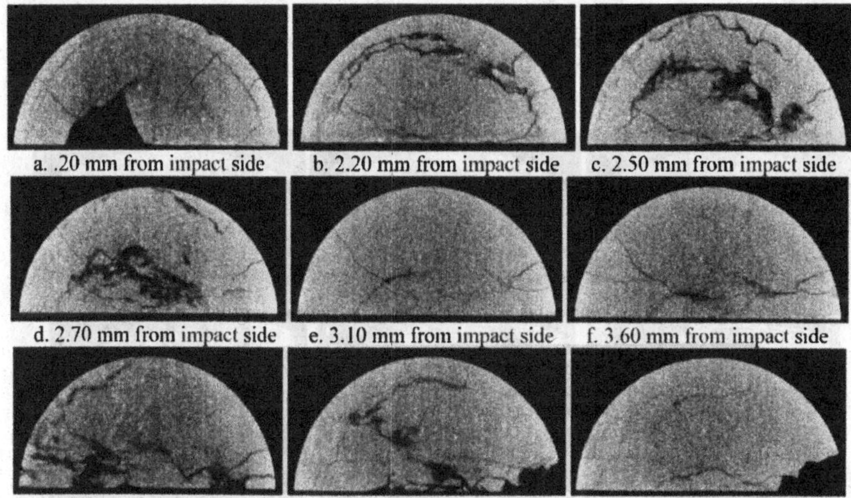

a. .20 mm from impact side b. 2.20 mm from impact side c. 2.50 mm from impact side
d. 2.70 mm from impact side e. 3.10 mm from impact side f. 3.60 mm from impact side
g. 4.30 mm from impact side h. 4.60 mm from impact side i. 5.10 mm from impact side

Figure 1. A series of XCT scans of the AD995 specimen out of its brass cup (surround).

3-D Volume Visualization

The excellent dimensional accuracy and the digital nature of XCT images allow the accurate volume reconstruction of multiple adjacent slices. A virtual 3-D solid image is created by electronically stacking each XCT image, which have thickness over their cross sections (i.e., voxels), one on top of the previous from the bottom to the top of the specimen to generate its virtual volume. The shades of black and white in the notches of missing material and the appearance of the surface cracks is due to the angle of the virtual lighting on the specimen. Figure 2 shows a series of 3-D solid images of the specimen with various sections virtually removed. The method of virtual sectioning, which is essentially only showing a portion of each XCT scan, allows viewing of generated surfaces anywhere in the volume of the specimen in 3-D space. The view in Figures 2.a, 2.b, and 2.c is looking slightly down at the flat side of the specimen with it tilted to see the impact side also. In Figure 2.a the virtual flat surface is physically just inside the actual sectioned surface of the specimen and the edge of the region of relatively severe damage closer to the rear side is evident by the presence of the very dark horizontal band in the flat surface. Similarly, Figures 2.b and 2.c, which incrementally shows less material perpendicular to the faces of the specimen, cut across the region of severe damage closer to the impact side as indicated by the higher dark horizontal lines. The surfaces in Figures 2.b and 2.c are about 1.0 mm and 2.0 mm from the actual sectioned surface, respectively. The lower region of severe damage is also evident in Figure 2.b. Figure 2.c also shows an interesting jagged crack like feature on the left between the upper and lower horizontal lines indicating damage along this path. Figures 2.d, 2.e, and 2.f have been rotated about the z (i.e., through thickness) axis in order to view a second virtual surface perpendicular and adjacent to the first. In these images the wider flat surface on the left is at the same location as in Figure 2.a and they incrementally show less material perpendicular to this surface. The distance to the surface on the right from the corner of the specimen is about 2.30 mm, 3.30 mm, and 4.40 mm in Figures 2.d, 2.e, and 2.f, respectively. The cracks in either of the surfaces can be followed into the adjacent surface. In Figure 2.f the two main cracks in the right hand surface are basically unconnected. Figures 2.g, 2.h, and 2.i are similar to Figures 2.d, 2.e, and 2.f, but show less of the specimen as indicated by less of the notch in the impact side being visible. Again, cracks in the surfaces can be followed from one into the other. In all three figures the left surface is about 1.90 mm from the actual sectioned surface. The distance to the surface on the right from the corner is about 2.30 mm, 4.40 mm, and 6.40 mm in Figures 2.g, 2.h, and 2.i, respectively.

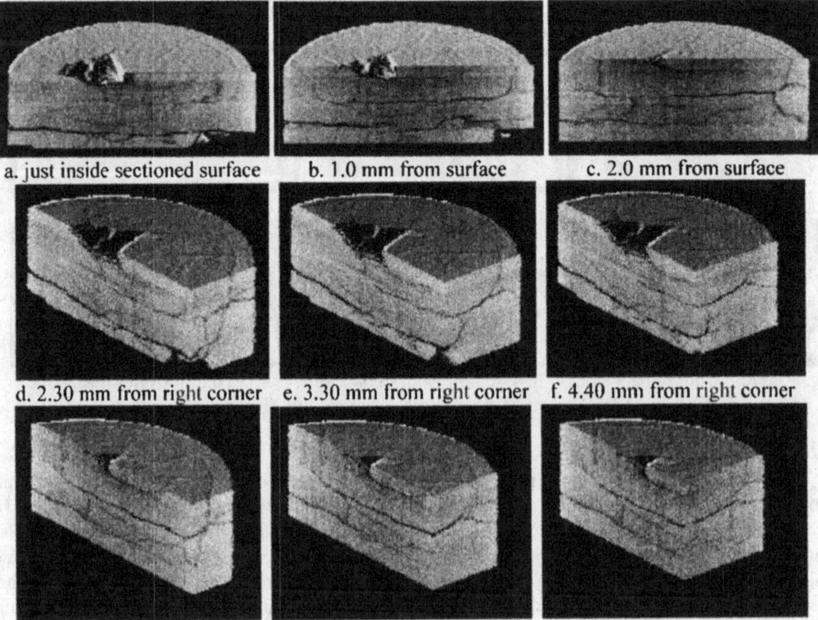

a. just inside sectioned surface b. 1.0 mm from surface c. 2.0 mm from surface

d. 2.30 mm from right corner e. 3.30 mm from right corner f. 4.40 mm from right corner

g. 2.30 mm from right corner h. 4.40 mm from right corner i. 6.40 mm from right corner

Figure 2. A series of virtual 3-D solid images of the AD995 specimen out of its brass cup showing various physical sections virtually removed. In (d-f) surface on left is just inside physically sectioned flat surface. In (g-i) surface on left is about 1.90 mm from physically sectioned surface. Images (g-i) appear thicker due to different processing (cropping) than images (d-f).

Quantitative Damage Results

Furthermore, the unit damage (decimal) fraction of a possible 100 percent (1.00) in the specimen was plotted as a function of depth (distance from the impact side of the specimen – face not directly impacted) for different segmentation threshold values. Figure 3 shows the effects of varying this threshold value for a given XCT slice in the specimen. At the top of Figure 3 is the normal distributed gray level slice that the thresholding and segmentation (i.e., binarization) were performed on. Below it are four binary damage images for different threshold values in which white indicates nominal ceramic material and black indicates damaged ceramic material within the specimen. Changing the threshold changes the relative levels of nominal and damaged material. Figure 4 shows the unit damage fraction plots for each threshold value together on the same graph. The damage fraction plots for the four lower threshold values, including 2350, are grouped relatively close together. However, the damage fraction plot for the highest threshold, 2500, is vertically separated from the others by a larger gap. This type of behavior can be indicative that the particular threshold is too high. Figure 3 shows that a certain physical amount of what is probably nominal undamaged material has been segmented to damaged material at the 2500 threshold. These observations indicate that the threshold of 2500 is not the best choice to generate a representative plot of damage fraction through the thickness of the AD995 specimen. More interesting though is the nature of the changes in damage fraction through the thickness of the specimen for a given threshold, since every plot trends in the same way, as expected. The maxima in the plots away from the impact (> 1.75 mm) and rear sides of the specimen correspond to the presence of significant planar damage due to a reflected tensile pulse thought to initiate at the rear side of the specimen.

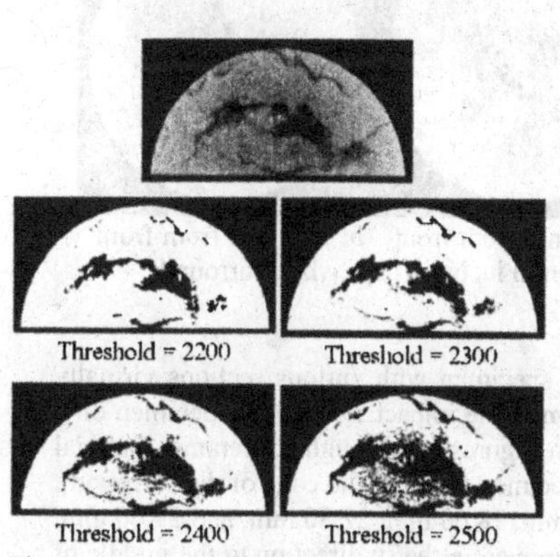

Figure 3. A normal distributed gray level XCT slice of the AD995 specimen and four binary damage images using different gray level thresholds.

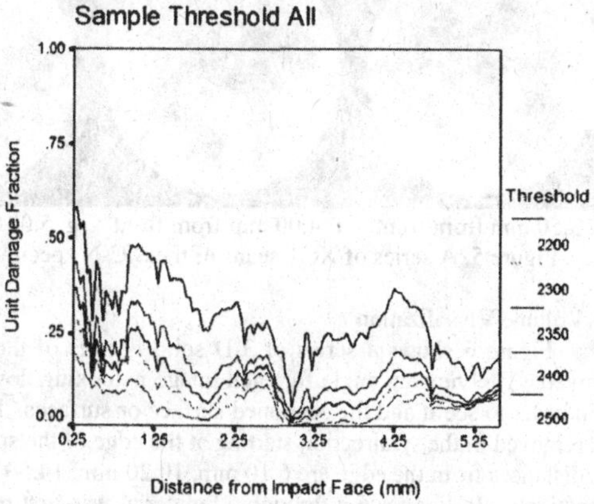

Sample Threshold All

Unit Damage Fraction: 1.00, .75, .50, .25, 0.00

Distance from Impact Face (mm): 0.25, 1.25, 2.25, 3.25, 4.25, 5.25

Threshold: 2200, 2300, 2350, 2400, 2500

Figure 4. Plots of Unit Damage Fraction vs. Slice Distance from Impact Side of the AD995 specimen as a function of gray level threshold value.

EVALUATION OF SiC SPECIMEN

XCT Results

Figure 5 shows a series of slices of the specimen starting at near the impact side and ending at near the rear side. The distances of the slices from the front face of the specimen are .50, 1.00, 2.00,

2.50, 3.50, 4.00, 5.00, and 5.50 mm in 5.a to 5.h, respectively. The first few images show radial cracks between the line of attack, which is slightly off center, and the relatively large circumferential, or ring, damage towards the outside of the specimen. This damage trend continues as the ring cracking closes in and decreases in severity going towards the rear of the specimen, at which a significant amount of material in the center is damaged or missing (Figure 5.h). The two faint darker rings at the center of some of the images are artifacts (i.e., not representative of actual physical features). Relatively isolated lateral cracking is present in the specimen 3.50 mm and 4.00 mm from its front face (Figures 5.e and 5.f), which becomes both longer and wider and appears to mix with radial cracking (left hand side) going towards the rear of the specimen (Figure 5.g).

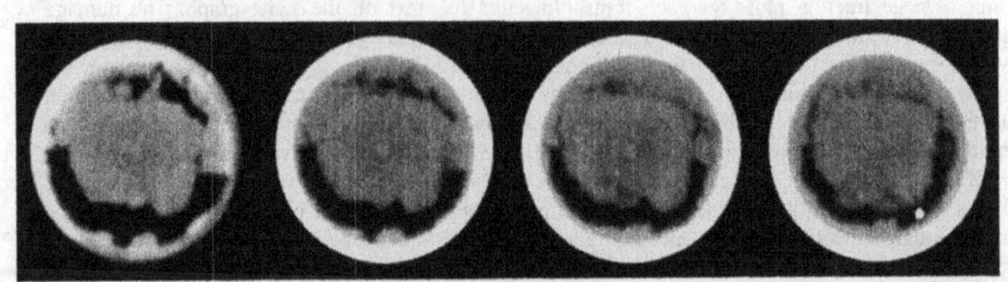

a. .50 mm from front face b. 1.00 mm from front c. 2.00 mm form front d. 2.50 mm from front

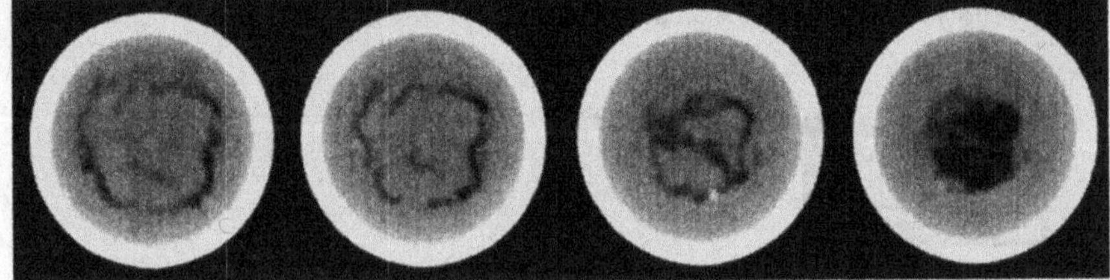

e. 3.50 mm from front f. 4.00 mm from front g. 5.00 mm from front h. 5.50 mm from front

Figure 5. A series of XCT scans of the SiC-N specimen in its brass cup (white surround).

3-D Volume Visualization

Figure 6 shows a series of 3-D solid images of the specimen with various sections virtually removed. The viewing angle for each image is looking down at the impact side of the specimen on a tilt in order to see it and the sectioned surface or surfaces. In Figures 6.a through 6.f ceramic material was removed in the -y direction starting at the edge of the specimen (i.e., not the edge of the surround). The distances from the edge are 6.10 mm, 10.20 mm, 14.50 mm, 18.60 mm, 22.80 mm, and 29.00 mm, respectively. In Figures 6.g through 6.l material was first removed in the -y direction to the middle of the specimen and then in the orthogonal +x direction again starting at the edge of the specimen. Figures 6.g through 6.l have also been rotated about the z-axis (i.e., through thickness direction) in order to view both of the perpendicular virtual surfaces. The distances from the edge of the specimen to the surfaces on the right are 4.20 mm, 8.30 mm, 12.60 mm, 17.60 mm, 21.80 mm, and 25.90 mm, respectively. The shape of the ring, or cone, cracking and the relatively severe damage near the impact side of the specimen, as well as radial cracking in some instances (Figures 6.i-6.k), are readily visible in the solid images. The shape of the cone crack is visible in Figures 6.b through 6.f and the crack can be followed from one surface into the adjacent surface in Figures 6.g through 6.l. The solid white banding at the top of the specimen in some of the images is an artifact due to a combination of virtual sectioning and lighting effects.

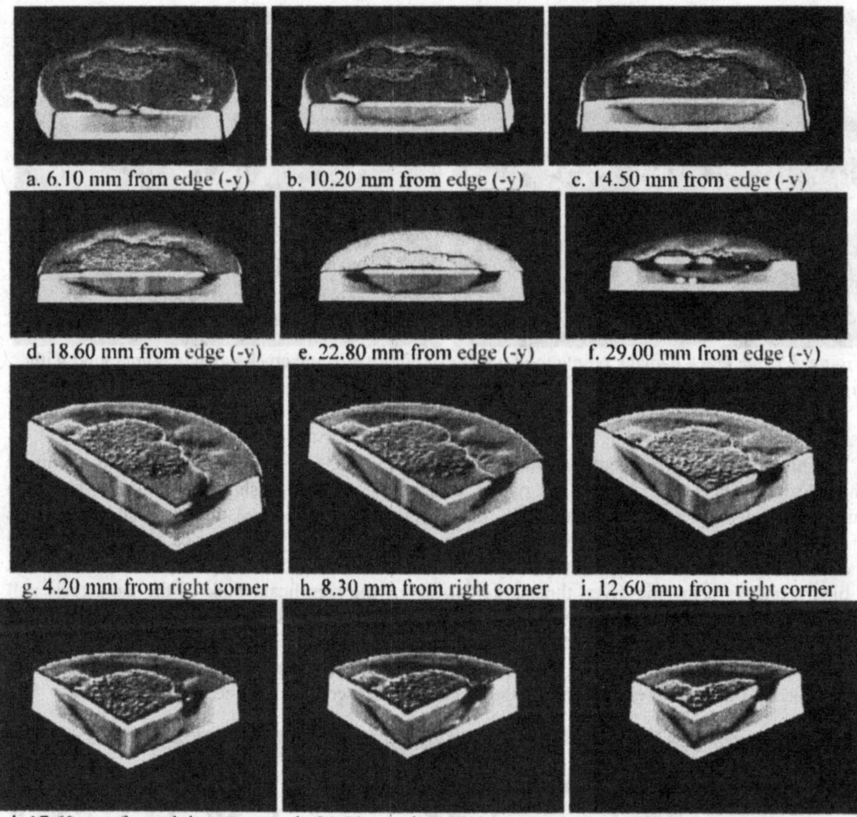

j. 17.60 mm from right corner k. 21.80 mm from right corner l. 25.90 mm from right corner

Figure 6. A series of virtual 3-D solid images of the SiC-N specimen in its brass cup (bottom and sides white surround) showing various physical sections virtually removed. In (g-l) half of the specimen has been virtually removed in the –y direction (i.e., the surface on the left).

The cone crack damage was isolated from the rest of the damage in the form of a 3-D point cloud in which the points in space define the edges of the damage, slice by slice, as shown in Figure 7. The viewing angle is also looking down at the impact side of the specimen on a tilt A free form 3-D surface, which is the wavy looking surface going through the points, fit to the damage point cloud is also shown in the same space with the relative location of the edges of the specimen delineated (top and bottom circles). Figure 7 also shows a line passing through the specimen delineating the angle of attack as determined from damage in the individual XCT slices. Figure 8 shows a simpler and more directly informative fit of a cone to the damage cloud the angle of which is 68.7 degrees, also with the relative location of the edges of the specimen delineated. The surface of the cone is defined by the thicker angled lines passing through the damage cloud. The cone subtends a total angle of 137.4 degrees from one side to the other.

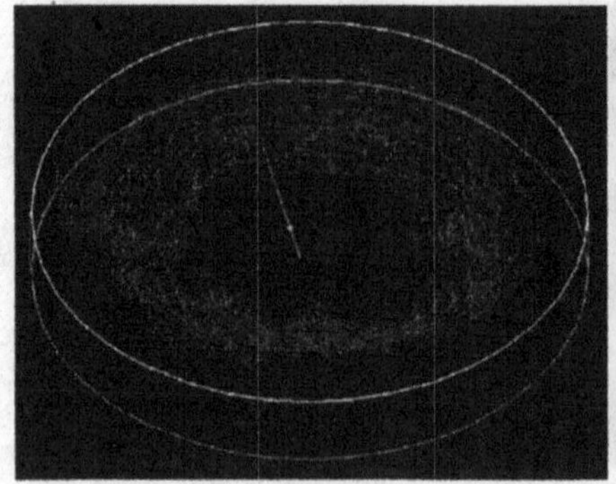

Figure 7. Cone crack damage point cloud with free form 3-D surface fit.

Figure 8. Cone crack damage point cloud with cone fit.

Quantitative Damage Results

Thresholding and segmentation were performed on the XCT slices of the specimen in similar fashion to the AD995 specimen. A set of resulting binary images are shown in Figure 9. in which damaged or missing ceramic material in the specimen is black and the surrounding cup is indistinguishable from the undamaged ceramic because they are both white. As can be seen in Figure 5 the radial cracks are faint compared to the rest of the damage, which is why they cannot be seen in the binary images using the given threshold. However, the significant bulk of the damage is still represented and provides a good quantitative measure of the level of damage. A relatively small amount of damage is visible as roughly circular black areas near the middle of the specimen in Figures 9.b and 9.c. Points in the damage cloud from these particular features in individual slices were mathematically fit to find their centers and the resulting data used to determine the line of attack

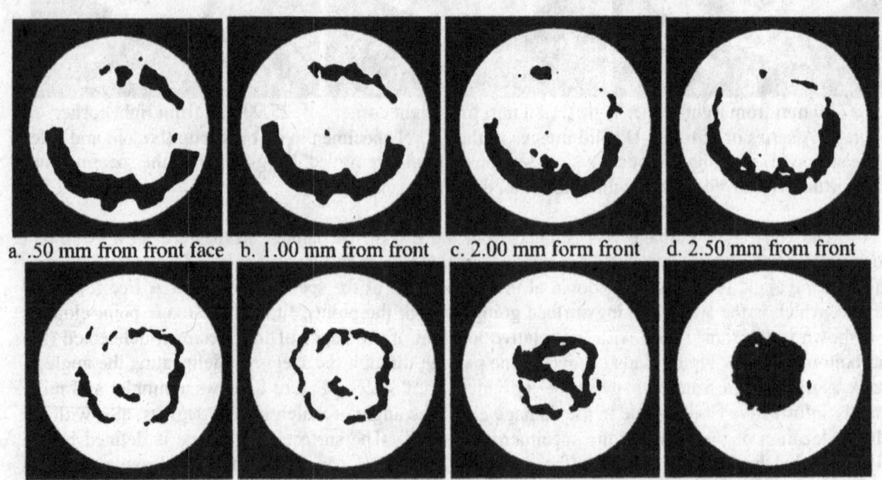

a. .50 mm from front face b. 1.00 mm from front c. 2.00 mm form front d. 2.50 mm from front

e. 3.50 mm from front f. 4.00 mm from front g. 5.00 mm from front h. 5.50 mm from front

Figure 9. A series of binary images of the XCT scans of the SiC-N specimen.

(Figures 7 and 8). Figure 10 is a plot of unit damage fraction over the entire area of the specimen as a function of depth with the surround removed from the calculations, which shows that the damage first decreases going away from the impact side of the specimen and then increases towards the rear of the specimen. The relatively large change in unit damage fraction from a depth of 2.50 mm to a depth of 3.00 mm is not understood at this time.

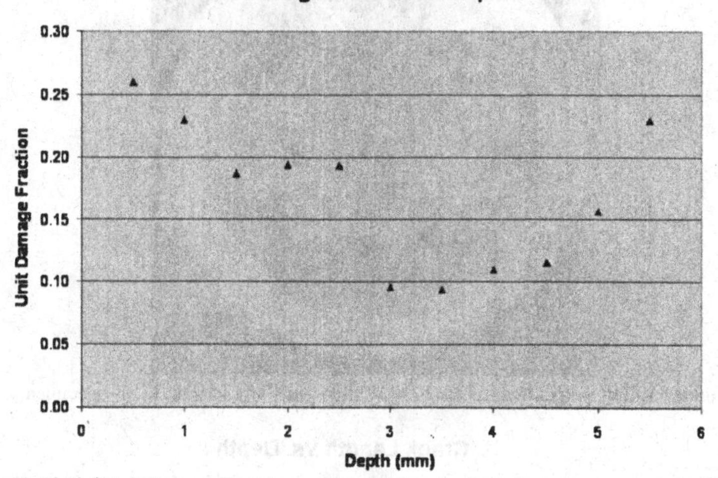

Figure 10. Plot of Unit Damage Fraction vs. Depth (distance from front face) over area of SiC-N specimen.

Figure 11 highlights five cracks of interest using black dashed lines between the line of attack and the outer circumferential, or ring, type damage in an XCT slice near the impact side of the specimen. The paths of the cracks are fairly well delineated and retain their individual nature through most of the thickness of the specimen with well defined start (line of attack) and end (ring damage) points. This was not the general behavior of the radial and lateral cracks through the thickness of the AD995 alumina specimen. The endpoints of the cracks were determined throughout the thickness of the specimen from the individual slices and their lengths subsequently calculated. Figure 12 is a plot of the length (endpoint to endpoint) of these particular cracks as a function of depth. In a few cases the endpoints of cracks at depths of 2.00 mm and 2.50 mm could not be reliably determined due to damage features resulting in some cracks missing one or two length data points. In the case of Crack 5, which was towards relatively less severe ring and cone cracking, it became very faint and difficult to follow, if not impossible, with increasing depth. It can be seen that the lengths of Crack 2 (solid diamonds) and Crack 4 (clear squares) follow a relatively smooth decreasing trend. However, the lengths of Crack 1 (solid triangles) and Crack 3 (clear circles) start off decreasing or remaining approximately the same with increasing depth and then increase at a depth of about 2.50 mm, although the length of Crack 3 increases to a lesser extent, after which their lengths follow a decreasing trend again. This interesting behavior occurs in an approximate location within the specimen where there is relatively more severe ring and cone cracking as well as lateral cracking (Figures 5.d through 5.f). The behavior may also be related to the nature of the changes in unit damage fraction with depth in that location.

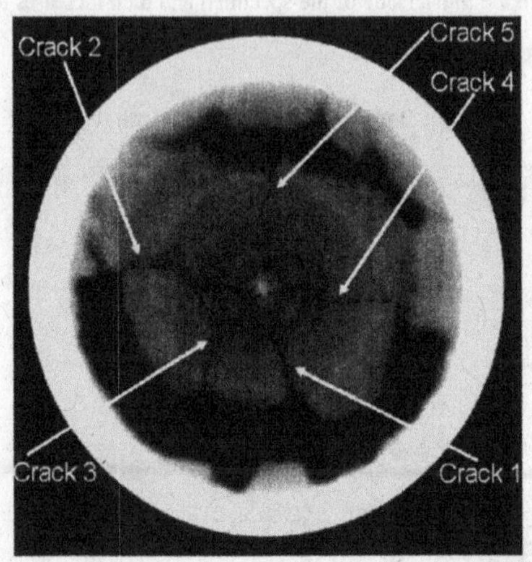

Figure 11. Five cracks highlighted in XCT slice near front face of SiC-N specimen.

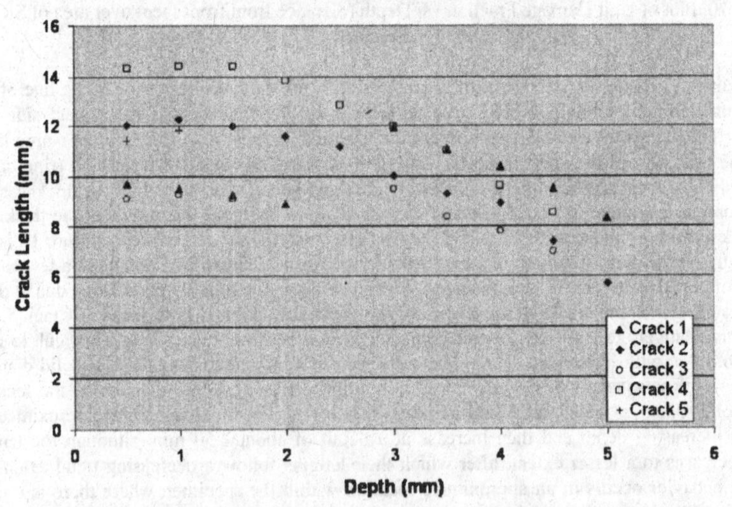

Figure 12. Plot of Length vs. Depth for a set of five individual cracks (Figure 11).

CONCLUSION

The shock induced damage in AD995 alumina and SiC-N ceramic targets was inspected and extensively evaluated using XCT 2-D and 3-D analysis. The shock damage in the AD995 alumina and SiC-N targets was induced by 6 GPa and 4 GPa pressures, respectively. Several types of damage features, including radial, circumferential or ring, cone, and lateral cracking were found in the specimens and discussed. Examples of quantitative analyses were also given and discussed, including unit damage fraction, cone angle, and crack lengths. Quantitative analyses could also include axi-symmetric (i.e., dependent on radius) – depth multivariate approaches. The methods discussed in this paper and other methods as well [1-17] provide effective means to obtain a quantitative level of data on shock and impact damage. The second part of the technical focus is how to utilize quantitative data in these forms to validate and improve or modify ballistic damage models. It is highly desirable to continue to develop the capabilities of the XCT diagnostic approach and apply results from it to ballistic damage modeling, creating an efficient, useful, and powerful investigative methodology for characterizing, analyzing, and comparative modeling of volumetric shock/impact damage. The overall goal of this work is the further critical development of the non-invasive/non-destructive XCT diagnostic approach to create a very efficient, useful, and powerful investigative methodology for characterizing, analyzing, and comparative modeling of volumetric shock and impact damage.

FOOTNOTES

*Work performed with support by an appointment to the Research Participation Program at the U.S. ARL administered by the Oak Ridge Institute for Science and Education through an interagency agreement between the U.S. Department of Energy and U.S. ARL.

REFERENCES

[1]W. Green, N. Rupert, and J. Wells, Inroads in the Non-invasive Diagnostics of Ballistic Impact Damage, *Proceedings of 25th Army Science Conference*, (2006).

[2]H. Miller, W. Green, and J. LaSalvia, Ballistically-Induced Damage in Ceramic Targets as Revealed by X-ray Computed Tomography, *Proceedings of 31st International Conference on Advanced Ceramics and Composites - Topics in Ceramic Armor*, (2007).

[3]N. Bourne, W. Green, and D. Dandekar, On the One-dimensional Recovery and Microstructural Evaluation of Shocked Alumina, *Proceedings of the Royal Society A: Mathematical, Physical, and Engineering Sciences*, published online: doi:10.1098/rspa.2006.1713, (2006).

[4]J. Wells, W. Green, N. Rupert, and D. MacKenzie, Capturing Ballistic Damage as a Function of Impact Velocity in SiC-N Ceramic Targets, *Proceedings of 30th International Conference on Advanced Ceramics and Composites - Advances in Ceramic Armor*, (2006).

[5]J. Wells, On the Role of Impact Damage in Armor Ceramic Performance, *Proceedings of 30th International Conference on Advanced Ceramics and Composites - Advances in Ceramic Armor*, (2006).

[6]J. Wells, Progress on the NDE Characterization of Impact Damage in Armor Materials, *Proceedings of 22nd International Ballistics Symposium*, ADPA, **2**, 793-800 (2005).

[7]W. Green, N. Rupert, and J. Wells, Investigating Ballistic Impact Damage in Lightweight Ceramic Armor Materials Using Advanced Computed Tomography, *Nondestructive Characterization of Materials XI*, 251-259, (2003).

[8]H. Miller, W. Green, N. Rupert, and J. Wells, Quantitative Evaluation of Damage and Residual Penetrator Material in Impacted TiB_2 Targets Using X-Ray Computed Tomography, *Proceedings of 21st International Ballistics Symposium*, ADPA, **1**, 153-159 (2004).

[9]J. Wells, W. Green, N. Rupert, J. Winter, J. Wheeler, S. Cimpoeru, and A. Zibarov, Ballistic Damage Visualization & Quantification in Monolithic Ti-6Al-4V With X-ray Computed Tomography, *Proceedings of 21st International Ballistics Symposium*, **1**, 371-377 (2004).

[10]J. Wells, W. Green, and N. Rupert, Nondestructive 2-D and 3-D Visualization of Interface Defeat Based Ballistic Impact Damage in a TiC Ceramic Target Disk, *Proceedings of 21st Army Science Conference*, (2002).
[11]W. Green, K. Doherty, N. Rupert, and J. Wells, Damage Assessment in TiB_2 Ceramic Armor Targets: Part I - X-ray CT and SEM Analyses, *Proceedings of 2nd International Conference on Mechanics of Structures (MSMS2001)*, (2001).
[12]N. Rupert, W. Green, K. Doherty, and J. Wells, Damage Assessment in TiB_2 Ceramic Armor Targets: Part II - Radial Cracking, *Proceedings of 2nd International Conference on Mechanics of Structures (MSMS2001)*, (2001).
[13]W. Green and J. Wells, Characterization of Impact Damage in Metallic/Nonmetallic Composites Using X-Ray Computed Tomography Imaging, *Nondestructive Characterization of Materials IX*, AIP, 622-629 (1999).
[14]W. Green, J. Wells, and N. Rupert, Comprehensive Visualization of Interface Defeat-Based Ballistic Impact Damage in a Titanium Carbide (TiC) Ceramic Target Disk, *Technical Report ARL-TR-2565*, U.S. Army Research Laboratory, (2001).
[15]J. Wells, W. Green, N. Rupert; J. Winter, and S. Cimpoeru, On the 3D Visualization of Ballistic Damage in Ti-6Al-4V Applique Armour with X-Ray Computed Tomography, *Proceedings of 22nd International Ballistics Symposium*, (2005).
[16]N. Rupert, J. Wells, W. Bruchey, and J. Wheeler, The Evolution and Application of Asymmetrical Image Filters for Quantitative XCT Analysis, *Proceedings of 23rd International Ballistics Symposium*, ADPA, **2**, 945-952 (2007).
[17]J. Wheeler, W. Green, M. Schuresko, and M. Lowery, A Framework for the Analyses and Visualization of X-Ray Computed Tomography Image Data Using a Compute Cluster, *Proceedings of 22nd International Ballistics Symposium*, (2005).
[18]R. Brannon, J. Wells, and O. Strack, Validating Theories for Brittle Damage, *Metallurgical and Materials Transactions A*, **38A**, (2007).
[19]J. Wells and R. Brannon, Advances in X-Ray Computed Tomography Diagnostics of Ballistic Impact Damage, *Metallurgical and Materials Transactions A*, **38A**, 2944-2949 (2007).
[20]K. Iyer and D. Dandekar, Computational Design Study for Recovery of Shock Damaged Silicon Carbide, *Proceedings of Shock Compression of Condensed Matter-2005*, AIP, 866-869 (2006).

AUTOMATED NON-DESTRUCTIVE EVALUATION SYSTEM FOR HARD ARMOR PROTECTIVE INSERTS OF BODY ARMOR

Nicholas Haynes
Project Manager Soldier Equipment
Clifton, VA USA

Karl Masters
Project Manager Soldier Equipment
Culpeper, VA USA

Chris Perritt
Project Manager Soldier Equipment
Broad Run, VA USA

David Simmons
Project Manager Soldier Equipment
Alexandria, VA USA

Dr. James Zheng
Project Manager Soldier Equipment
Manassas, VA USA

Dr. James E. Youngberg
JDLL, Inc.
Utah, USA

ABSTRACT

To properly protect personnel, the ceramic plate component in body armor must be free from cracks. Studies by the US Army and Britain[2] have shown that while cracked plates still can defeat a threat, their ballistic performance is degraded. The Non-Destructive Automated Inspection System (NDE-AIS) is a deployable, high speed, automated, digital radiographic inspection system that evaluates ceramic plate serviceability in the field. During the most recent assessment, digital x-rays of plates were examined by the NDE-AIS and the results were compared with three radiographer's assessments of the same plates. The results demonstrated that the NDE-AIS is 99.8% effective in keeping defective body armor plates from being reissued to Soldiers. Ceramic plates are inspected at an average rate of about 240 plates per hour in a process that automatically identifies and withdraws defective plates from service. Inspection results are maintained in a database along with CAGE codes (manufacturer), NSNs (make and size), lot numbers, serial numbers and dates of manufacturing information. This information can be use to help facilitate inventory, logistical activities and production analysis. The mission, design, and effectiveness of the inspection system are discussed herein.

INTRODUCTION

Current designs of body armor normally incorporate ceramic tile and high performance fiber composite hard inserts to provide rifle protection for an individual. Incorporated into the design are many features that make each design unique even though it may protect against the same threat level. However, most designs utilize a ceramic tile strike face to provide superior ballistic performance at a light weight. Some of the common ceramics used in body armor are aluminum oxide, silicon carbide, aluminum nitride and boron carbide[4]. The ceramics are manufactured differently which gives the material different characteristics. Reaction bonded and hot pressed ceramics are some of the common manufacturing processes. The material and manufacturing process used to form the ceramic is carefully considered in an effort to reach an optimal tradeoff between ballistic performance and durability. While ceramics are very hard, they are also brittle. Ceramic tiles are designed and manufactured to withstand certain loads and impacts, but even the toughest plate is susceptible to cracking. There can be several causes for this, but in most cases it is a result of mishandling or abusive use. It is extremely important to identify a cracked plate because the item may fail when it is needed most[2].

The Army currently uses a "torque test" in the field to evaluate hard armor plates. This is a process in which an individual grabs opposing corners of a plate and tries to twist the plate listening for crunching or cracking. When this test works, adjoining surfaces of a crack rub and create the sound that reveals the crack. Unfortunately, this easily-performed field test is not always reliable.

Non-destructive testing (NDT) methods that may be applicable to this application include radiography, ultrasonics[1], thermal imaging[3], computed tomography (CT scan) and eddy current testing. Other researched test methods include electrical properties profiling and sonic resonance. Each method has advantages and disadvantages, and no single test method will detect every flaw. The "best" method optimizes performance with respect to the unique characteristics of the application. Digital radiography was ultimately chosen for this application because of (1) its capability to support high throughput, (2) its unique relative insensitivity to the considerable variations that exist in both the design, size and the condition of the insert and it's covering, (3) its ability to detect at least some portion of the cracking in virtually every cracked plate, and (4) its ability to produce an understandable image record of inspected plates for verification purposes.

In almost all of the NDT methods researched, the covering of the plate created problems when trying to evaluate the serviceability of the item. A large majority of the plates and all new plates have protective coverings consisting of foam or like items. In most cases the plate cover would have to be removed in order to evaluate the condition. This of course would compromise the plate and its ruggedness.

The Army has developed a Non-Destructive Evaluation Automated Inspection System (NDE-AIS) that automatically identifies plates, evaluates and separates serviceable and non-serviceable plates, while maintaining a high throughput. It also has the capability to query data to provide logistical support and increase product quality. This system is the first automated system to test ballistic integrity of body armor components without the need of a radiographer.

DESCRIPTION

There are several subsystems that comprise the NDE-AIS. The first is the shell of the system which contains the Automated Inspection System and makes the system mobile. The shell is the Hard sided Expandable Light Air Mobile Shelter (HELAMS) and is shown in Figure 1. The HELAMS has a detachable wheel set and can either be hauled or airlifted to a working location. It can sustain itself with a diesel generator. but it can also be connected to shore power. The HELAMS is climate controlled and creates a comfortable working environment for the user.

Figure 1: HELAMS with front wheel set detached.

Within the HELAMS is the Automated Inspection System (AIS). The AIS consists of a conveyor. x-ray source. x-ray imager and computer with software required to analyze the plate images. A 3-D rendering is shown in Figure 2 and the system in use is shown in Figure 3.

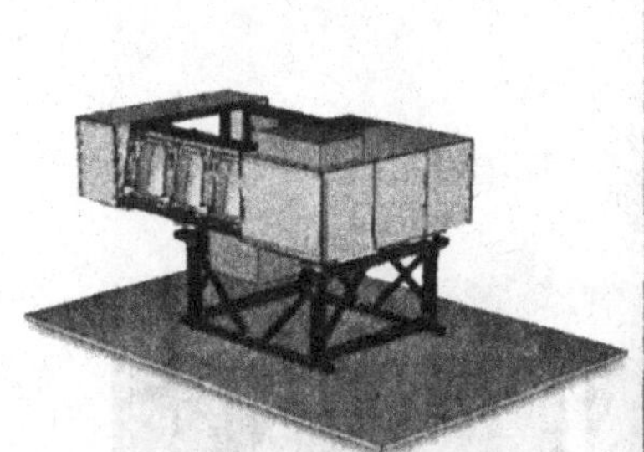

Figure 2: 3-D rendering of AIS.

Figure 3: NDE-AIS in use during field test.

The NDE-AIS examines over 240 plates per hour. The only interaction required of the user is loading and unloading the plates while controlling the system with a touch sensitive screen. Acceptable plates continue on the conveyor belt and can be offloaded on the opposite side of the machine. Defective plates are automatically identified and dropped into a reject bin to separate them from the serviceable plates.

The AIS performs fully-automated x-ray, imaging acquisition and analysis of plates presented to the system. Analysis works by separating features expected in each x-ray image – those due to the ceramic plate, its backing, and cover – from features that reveal plate cracking. The algorithms that do this are designed so that the fraction of damaged plates mistakenly accepted can be traded against the fraction of undamaged plates that are rejected. The algorithm and the hardware on which it runs (a quartet of 64-bit AMD Opteron CPUs) are optimized to support the AIS's four-plate-per-minute average throughput. Finally, to the extent feasible, the data analysis has been designed to accommodate plate design variations with minimal software adjustments.

The challenge of removing expected projection features is complicated by their number and diversity. The algorithm accounts for variables due to radiographic and imager processes such as exposure, varying x-ray field shape, plate-dependent x-ray scatter, and the imager's sensitivity field (its constant and time-varying components). It also accounts for a variety of plate sizes and designs, as well as plate-to-plate variations in fabrication technique that occur due to differences among factory personnel. Among the most challenging imaging artifacts is the fine local variation that arises as a combination of imager noise, photon noise, and as a result of the arrangement of fibers in the plate backing and cover. In many cases, crack indications are actually smaller than the level of interference imposed by these fine local variations, necessitating the use of algorithms related to the Hough transform. Our overall strategy, which accomplishes the above-described processing, is comprised of the following steps:

1. Remove artifacts of measurement
2. Remove expected structure
3. Minimize the effects of measurement noise
4. Identify and describe areas having related pixels
5. Remove noise areas
6. Remove any areas reflecting expected structure
7. Test remaining areas against criteria

Results of various automated processing steps are displayed in Figure4.

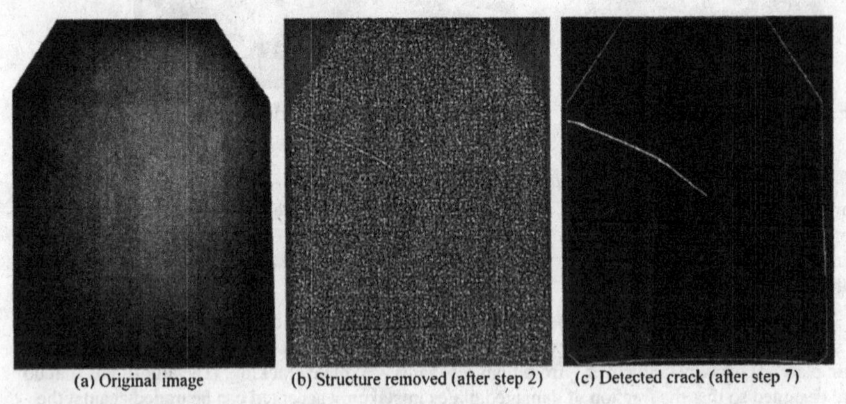

(a) Original image (b) Structure removed (after step 2) (c) Detected crack (after step 7)

Figure 4: Example of crack detection processing (referencing processing steps from the text).

The x-ray equipment is composed of an x-ray source, which generates the x-rays and an imager which captures the x-ray image electronically. Because the behavior of the x-ray source and the imager can change over time, means are provided whereby these components can be calibrated. Calibration is fully automatic and is initiated by configuring a plate holder to present an empty field to the imager. Calibration is normally performed shortly after AIS startup on a daily basis. Individual algorithm steps are controlled and tuned using parameters maintained in a system configuration file. These parameters can be used to improve system accuracy as an increasingly large collection of plates with known results becomes available.

As each plate enters the AIS imaging area its UID is read. UID data accompany the plate throughout the AIS inspection and are entered in fields in the AIS database. The UID is also used to construct the file name of the temporary result image that is maintained temporarily in a first-in-first-out (FIFO) method for possible use in offline auditing. In addition to the UID information, various system, plate, and evaluation parameters are stored in the tables of the AIS relational database. These parameters provide information useful for system maintenance, especially for subsequent logistical analysis. An example of the data storage means is shown in Figure 5.

Plate ID with Serial Number	Cage Code (Manufacturer)	Lot Number	NSN (Make and Size)	Status
52969_E004R_297693	52969	E004R	8470-01-520-7373	accept
52969_E004R_297700	52969	E004R	8470-01-520-7373	accept
52969_E004R_297708	52969	E004R	8470-01-520-7373	accept
52969_E004R_297714	52969	E004R	8470-01-520-7373	accept
52969_E005_287330	52969	E005	8470-01-520-7373	accept
52969_E005_287332	52969	E005	8470-01-520-7373	accept
52969_E005_287356	52969	E005	8470-01-520-7373	accept
52969_E005_287387	52969	E005	8470-01-520-7373	accept
52969_E005_287388	52969	E005	8470-01-520-7373	accept
52969_E005_287409	52969	E005	8470-01-520-7373	reject
52969_E008_290973	52969	E008	8470-01-520-7373	accept
52969_E008_290975	52969	E008	8470-01-520-7373	accept
52969_E008_290979	52969	E008	8470-01-520-7373	accept
52969_E008_290980	52969	E008	8470-01-520-7373	accept
52969_E008_290986	52969	E008	8470-01-520-7373	accept
52969_E008_290989	52969	E008	8470-01-520-7373	accept
52969_E008_290991	52969	E008	8470-01-520-7373	accept

Figure 5: Sample of NDE-AIS database

To validate the concept and effectiveness of the NDE-AIS, two operational assessments were conducted using control samples in January 2007 and August 2007. These demonstrations also helped validate the operation and throughput of the NDE-AIS. Over 4,700 ceramic plates were inspected during these operational assessments.

The plates were grouped into batches of plates and loaded onto the roller assembly of the loading dock of the HELAMS. UID labels created by a laser labeling system were placed on the plate's back lower side to be read by the UID reader on the AIS system. Next, the plates were hand evaluated using the Army's "torque test". Each ceramic plate's condition was noted and then placed on the conveyor to be evaluated by the NDE-AIS. In order to test the accuracy of the system, a blind manual review of each x-ray image in the NDE-AIS database was performed by several certified radiographers. Separately, a hardware check was conducted with 50 plates from each batch using a separate manual Vidisco x-ray system to ensure that the radiographic hardware within the AIS was capturing an accurate image. Once all the plates were evaluated by the NDE-AIS. the plates were segregated into accept and reject piles.

In the first operational assessment, there was one radiographer doing the blind manual review to grade the NDE-AIS. In the second operation assessment, there were three radiographers who evaluated the results. After each radiographer reviewed images, their results were recorded and compared. If there was a difference in any of the three conclusions (accepted, rejected or possible damage) the three radiographers reviewed the appropriate image and came to a conclusion. Once a consensus was achieved, the AIS was judged upon the results of the radiographers which were considered ground truth. In any case the ground truth was either accepted or rejected. Possible damage could not be a final result.

RESULTS

The first operational assessment was very productive. Over 2000 plates were evaluated including Small Arms Protective Inserts (SAPI) and Enhanced Small Arms Protective Inserts (ESAPI). Figure 6: 6 summarizes the overall accuracy of the NDE-AIS in evaluating all 2,166 plates. The most critical value of the NDE-AIS is the false accept rate (plates which the NDE-AIS accepted but the radiographer found unserviceable) which is referenced as alpha. After the analysis. this was found to be the lowest percentage of the plates at 3% (64 plates). False rejects, which consist of plates the NDE-AIS rejected but the radiographer found serviceable (referenced as beta), consisted of 10% of the plates. The beta value represents the cost to the user with the disposal of possibly good plates. However these plates could also be redirected to be used for training purposes only. The true accept and reject percentages are green. The radiographer and NDE-AIS were in agreement a large majority of time (87%).

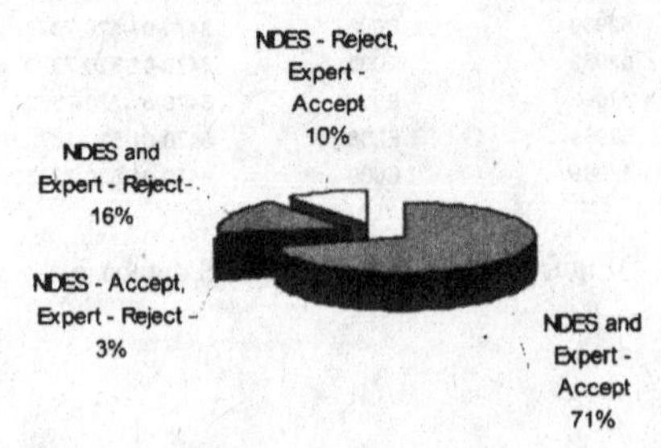

Figure 6: NDE-AIS Operation Assessment #1 Results.

After the conclusion of the first operational assessment, work continued on evaluating the plates that were categorized in the alpha and beta regions. The evaluation software was updated to correct for the anomalies of the system. The second operational assessment had significantly better results and helped with the continuation of updating the software component of the AIS. There were a total of 2.661 plates evaluated and the results can be seen in Figure 7.

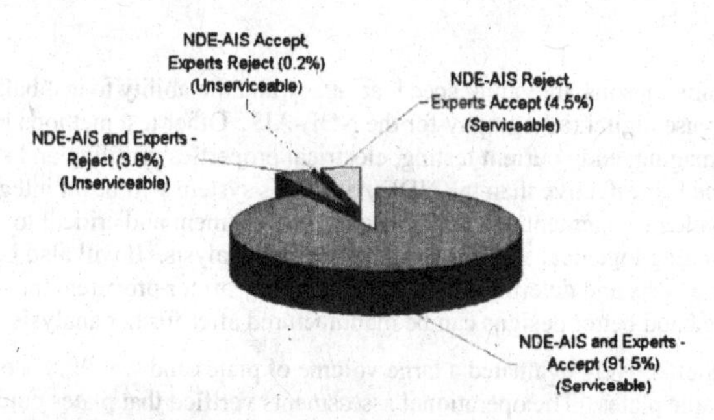

Figure 7: NDE-AIS Operation Assessment #2 Results.

The alpha value was improved dramatically as the rate was reduced from 3% in the first assessment to 0.2% which is shown in red in Figure 7. The beta value was also more than halved from 10% to 4.5% which is shown in yellow above. These are extremely favorable results especially with the reduction in the alpha value, which favors Soldier survivability.

DISCUSSION

The NDE-AIS algorithm can be adjusted to appropriately balance the alpha and beta percentages. These two values are inversely related. If a lower beta value is desired, the alpha value will increase. For example, opting for a decreased alpha value favors safety at the expense of increased operation cost due to unnecessary plate rejection. A key to reducing the amount of both parameters is by allowing the NDE-AIS to become familiar with internal features normal to the production characteristics of the plate by analyzing a radiographic image of a good plate. This is an important part in keeping the performance of the NDE-AIS as accurate as possible. Because the peculiarities of each design of each manufacturer are programmed into the software, part of the lifecycle management will be analyzing several plates of each newly approved design and size. This will allow the altering or adding of code to account for the design features.

It should be noted that with reference to the measured 4.5% beta value in the second operational assessment, the principal objective in the evaluation algorithm design was the minimization of the AIS alpha value– a choice in favor of user safety. Having achieved a reasonable performance in this respect, certain optimizations, planned for the near future, are available to reduce the beta value even further while keeping the alpha value at a minimum. As shown from the progress from the first to the second operational assessment, both values can be dramatically reduced with the analysis of more plates.

It was also observed that as the radiographers studied the images, naturally, fatigue and monotony lead to human error. There were some instances where a radiographer failed to identify a crack in the plate. This was discovered when the three evaluations of the radiographers were compared. It is also astonishing that the NDE-AIS evaluated the plates in one day. Although the radiographers had the help of the NDE-AIS stored images of each plate, they took approximately three full 8 hour days to evaluate the images. It should be noted that had the radiographers needed to manually x-ray each plate, the process may have taken several more days.

CONCLUSIONS

There were many reasons, including speed, accuracy and the ability to test ballistic plates that lead to the decision to use digital radiography for the NDE-AIS. Other test methods including ultrasonics, thermal imaging, eddy current testing, electrical properties profiling and sonic resonance have been proven to be less effective than the NDE-AIS. This system will be an integral part of determining the lifecycle management of the ballistic plate investment and critical to soldier survivability by facilitating logistical activities and production analysis. It will also be a key component in factoring costs and determining a budget for body armor procurement. Failure rates can accurately be measured and better designs can be manufactured after further analysis.

The NDE-AIS effectively evaluated a large volume of plates and was 99.8% effective in identifying unserviceable plates. The operational assessments verified that plates could be examined at a rate of over 240 plates per hour. This system has been designed for large scale projects and has been made mobile so that this scarce resource can be allocated efficiently to where it is needed. The HELAMS shelter affords the system this flexibility allowing transport by land, sea and air. The NDE-AIS could also become an integral system in hard body armor management for other services, agencies and law enforcement.

REFERENCES:

[1]Colanto, David, et al. "Automated Damage Assessment System For Ballistic Protective Inserts Using Low Frequency Ultrasonics." 2007.

[2]Dulay, Dee, et al. "Update on Practical Non Destructive Testing Methods for In-Service QA of Ceramic Body Armour Plates." PASS 2006 2006.

[3]Maresca, Joseph W., George L. Crawford, and Wilhelmina Dickerson. "Demonstration of an Engineering Prototype, including a Signal Processing Algorithm for Automation." Vista Engineering Technology 2 Nov. 2006.

[4]Tobin, Laurence B., and Michael J. Iremonger. Modern Body Armour and Helmets: An Introduction. Canberra, Australia: Argos Press, 2006.

[5]"A Rapid Non-Destructive Evaluation Inspection Technique Using Electrical Properties Profiling Method." Idaho National Laboratory, May 2006.

ANALYSIS OF HARDNESS INDENTATION SIZE EFFECT (ISE) CURVES IN CERAMICS: A NEW APPROACH TO DETERMINE PLASTICITY

Trevor E. Wilantewicz
ARL/ORAU Postdoctoral Research Fellow
U.S. Army Research Laboratory, APG, MD

James W. McCauley
U. S. Army Research Laboratory, APG. MD.

ABSTRACT

The hardness of a ceramic is generally believed to play a role in governing the ballistic performance of armor ceramics. Although the variation of hardness with load. i.e., the Indentation Size Effect (ISE), can make hardness comparisons between materials difficult, it may also provide more information on material behavior than hardness alone measured at a single load. In this work, several methods for curve fitting hardness-load data have been compared for both Knoop and Vickers hardness on several ceramic materials including aluminum oxynitride (AlON), silicon carbide, aluminum oxide and boron carbide. Whereas the Vickers hardness tended to reach a limiting plateau value with increasing load. the Knoop hardness did not. As a result of this, equations which predict a plateau value of hardness do not fit Knoop data well. Rather, a power-law equation ($H = kF^c$) is shown to fit the Knoop data much better. A plot of $\log_{10}$ (HK) vs. $\log_{10}$ (F) yielded easily comparable straight lines, whose slope and intercept data might be useful parameters to characterize the materials. It is proposed that the absolute value of the reciprocal of the slope is a measure of plasticity and that the sum of this value with the calculated Knoop hardness at 1 N may be a useful parameter to correlate to transitional velocity.

INTRODUCTION

In general, it has long been known that hardness correlates with gross ballistic performance, however, not with more subtle differences. Key properties and related laboratory scale tests that can correlate with ballistic armor performance have been elusive. Therefore, there is a need to identify quasi-static and dynamic laboratory scale tests, or ways of interpreting current test data, for materials development, analytical modeling, and materials screening purposes, rather than going to full blown ballistic tests early in the development of new materials. It is still not at all clear what combination of static and dynamic mechanical properties (figure of merit) control armor performance and how are these properties controlled/influenced by intrinsic (crystal structure, phase transitions, single crystal elasticity) and extrinsic material characteristics (composition/phase, grain level sub-structure, microstructure and processing defects).

Flinders et al.[1] found that for several silicon carbide materials the ballistic mass efficiency increased predictably with increasing Knoop hardness. However, no correlation was seen with the ballistic limit velocity (V_{50}) measured on the same materials, indicating that other properties combined with test configuration may play an important role as well.

As pointed out by Chheda et al.[2], hardness may be an indicator of a ceramic materials potential ballistic performance, and fracture toughness an indicator of how much of that potential can actually be attained. That is, the ceramic must remain intact as much as possible, in order to keep realizing the benefits of high hardness, i. e., high pressure on the projectile leading to erosion and fracture. Of course, the overall ceramic armor system design has as much to do with the performance as do the ceramic material properties.

Limited past success in relating quasi-static and dynamic mechanical property measurements to ballistic performance may have been due to using the wrong or incomplete properties or the lack of appropriate standardized tests that lacked reproducibility. Armor ceramic performance can be simplified into two main stages: a dwell phase, where the projectile velocity is nominally zero at the ceramic front face, and a penetration phase through fragmented ceramic. If the projectile is completely stopped at the front surface, this is referred to as "interface defeat". Lundberg et al.,[3] in classic work, suggested that the compressive yield strength modified by the amount of "dynamic plasticity" in armor ceramics correlates well to transitional velocities ("dwell"), that velocity (or impact pressure) where penetration begins. Note that "dynamic plasticity" is not a quasi-static fracture toughness, rather, it is metal-like property that refers to inelastic, non-linear deformation prior to catastrophic failure. Appropriately configured dynamic compression strength measurements in a Kolsky Bar, also referred to as a Split Hopkinson Pressure Bar (SHPB), have recently been suggested (Pickup and Barker, 1997)[4] as a way of determining the "dynamic plasticity". Work by James, 1996,[5] seemed to substantiate a possible correlation to ballistic performance. McCauley and Quinn, 2006, have recently reviewed this work. [6]

With respect to the ceramic material, if hardness, strength, elastic modulus, and fracture toughness were the only variables governing performance, then vast improvements in performance should be realized on a continuous basis. Since this is not the case, other properties, some of which may be unknown at this time, influence the performance in ways not completely understood. Conceptually, in hardness-load curves there are three main regions: a predominantly elastic region, a predominantly plastic region and a region of extensive permanent damage where extensive cracking is occurring. A fracture toughness value (K_{IC}) can be determined in the region of permanent damage, but this is not a measure of plasticity. The "hardness" is normally determined at a defined load and there are techniques and standard methods for doing so. The variation of indentation hardness with load[8-10], called the Indentation Size Effect (ISE), necessitates the comparison of material hardness at the same loads. Swab[11] and several ASTM standards[12,13] give guidelines for doing so.
Analysis of the entire indentation hardness-load curve may provide further insight into material behavior, especially a more careful analysis of the predominantly plastic region, and is the focus of the current paper. In particular, can appropriate parameters be extracted from load-hardness curves that would correlate with transitional velocity.

It will be shown that a power-law equation fits Knoop hardness-load curves better than previous fits[14], and may be an effective way to readily determine subtle differences in the ceramics.

EXPERIMENTAL PROCEDURES

In the previous paper[14], the Knoop data were fit using the following equations:

$$HK = (a/F) + b \quad (1)$$

where,

a,b = constants

F = indentation load

$$HK = (a_1'/d) + a_2' \quad (2)$$

where,

a_1', a_2' = constants

d = diagonal length

Regression analysis using these equations did not exhibit consistently high correlations to the data.

In the current work, the same data was fit using a power-law equation of the form:

$$HK = kF^c \quad (3)$$

where HK is the Knoop hardness (N/m^2), F indentation load (N), and k,c are constants determined by a computer regression analysis; c is dimensionless and k has unusual units of $N^{(1-c)}/m^2$. Taking the $\log_{10}$ of both sides of Eq. 1 yields:

$$\log_{10} HK = \log_{10} k + c \log_{10} F \quad (4)$$

If Eq.3 accurately describes the data, then a plot of $\log_{10}HK$ vs. $\log_{10}F$ will yield easily comparable straight lines with slope c. Knoop hardness data was also measured on two hot-pressed boron carbide materials, one from Ceradyne Inc. and another from what was formerly Cercom Inc, and was collected using the same instrument, and careful procedures, described in the previous paper[14]. In addition, the Knoop hardness of four hot-pressed silicon carbide materials was measured at 19.6 N and compared to the transition velocities measured in the reverse ballistic impact experiments conducted by Lundberg and Lundberg[15].

Vickers hardness data were also generated on the same specimens, using the same instrument and procedures. A 15 second hold time at the maximum load was used. For these data, in addition to using Eq. 3, an additional equation was utilized:

$$HV = \frac{a}{F} + b \quad (5)$$

where HV is the Vickers hardness (N/m^2), F indentation load (N), and a,b are constants determined by a computer regression analysis; b has units of N/m^2 and a has units of N^2/m^2. Although hardness can be plotted as a function of the diagonal size of the indentation, it is believed more appropriate to plot data as a function of load, since load is the independent variable that is under direct control. The constant 'a' gives the magnitude of the ISE, with larger values indicating a greater change in hardness for a given load interval.

RESULTS

Fig. 1 compares the power-law curve fit (Eq. 3) with the curve fit using Eq. 5 for the Knoop hardness of SiC-N. Note how Eq. 3 fits the data over the entire load range much better than Eq. 5, particularly at the higher loads, as shown by the much higher R^2 for the former than the latter i.e.,. 0.96 compared to only 0.67, respectively. The Knoop hardness data for all the materials is shown in Fig. 2, plotted using Eq. 4. The slope, c, is indicative of the rate of change of Knoop hardness with load. A negative value indicates the hardness decreases with load, as expected. The y-intercept hardness i.e., HK at $\log_{10}F = 0$, corresponds to the predicted Knoop hardness at F = 1 N, and which is very close to the hardness measured at F = 0.98 N. The slope and intercept data from the Knoop tests are summarized in Table I. Also shown in Table 1 are the R^2 values of the curve fits from Eq. 3, and those obtained from Eq. 5.

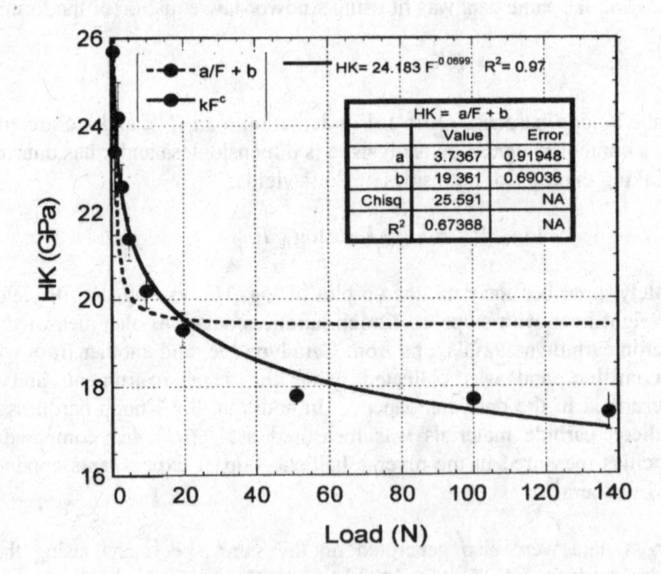

Fig. 1. Plot showing two curve fits to the Knoop hardness-load behavior of SiC-N. The power-law fit describes the data the best, as indicated by its greater R^2 value. Error bars are ± standard deviation of average values.

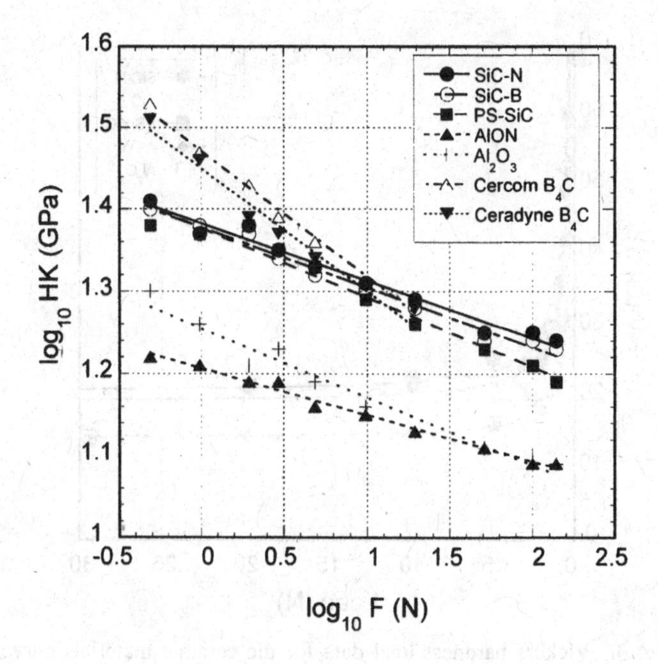

Fig. 2. Knoop hardness-load data for the ceramic materials curve fitted using Eq. 4.

Table I. Knoop Hardness Curve Fit Data

Material	Y-int. HK (GPa)	Slope, c	R^2	R^2 from (a/F + b) fit
SiC-N	24.1	-0.0699	0.97	0.67
SiC-B	23.8	-0.0728	0.98	0.67
PS-SiC*	23.8	-0.0845	0.96	0.51
AlON	16.0	-0.0565	0.98	0.65
AD995 CAP3 Al_2O_3	17.9	-0.0832	0.96	0.81
Cercom B_4C	29.6	-0.1536	0.98	0.90
Ceradyne B_4C	28.1	-0.1499	0.96	0.91

*PS-SiC = pressureless sintered SiC

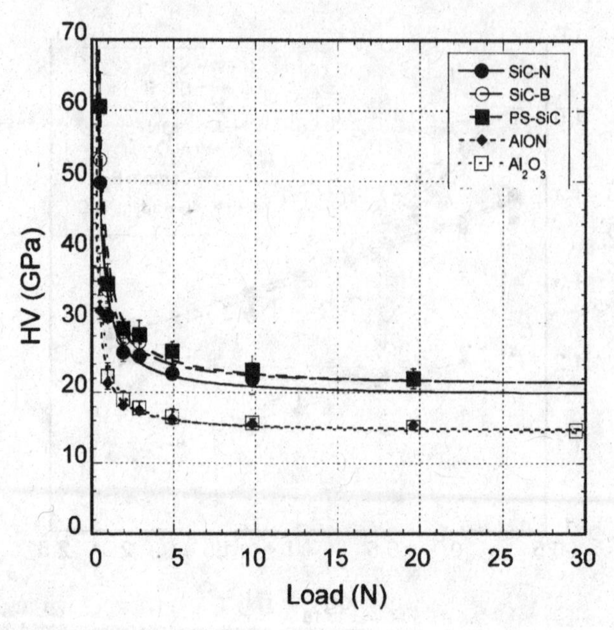

Fig. 3. Vickers hardness-load data for the ceramic materials curve fitted with the equation HV = a/F + b. (Eq. 5)

Fig. 3 illustrates the curves for the Vickers hardness as a function of load with Eq. 5 as the curve fit to the data. Table II summarizes the coefficients a, b and the R^2 values. The high R^2 values indicate good fits to the data. Excessive cracking in the SiC-N and SiC-B materials prevented the collection of reliable data at higher loads.

Table II. Vickers Hardness Curve Fit Data

Material	a (N^2/m^2) x10^9	b (N/m^2) x10^9	R^2
SiC-N	14.1	19.4	0.97
SiC-B	15.2	20.9	0.98
PS-SiC	18.4	20.7	0.97
AD995 CAP3 Al_2O_3	8.0	14.4	0.98
AlON	10.5	14.0	0.97

DISCUSSION

The Knoop hardness results show that the power-law equation (Eq. 3) fits the data much better than the equations based on a plateau value of hardness. From Table I it is seen that the rate of change of hardness, given by the slope c, is greatest for the boron carbide materials, despite the fact these materials have the highest hardness of all the materials at the lower loads. The relatively decent fit of Eq. 5 to the boron carbide data (see Table I) is a result of the rather limited load range examined, up to 20 N, compared to the other materials which were tested up to 137 N. Of the three silicon carbides, the pressureless sintered SiC has a higher slope compared to the two hot-pressed SiC materials. The AD995 CAP3 alumina has a higher slope compared to the hot-pressed SiC materials. Data for the AlON needs to be interpreted with caution, because of the large grain size of AlON (150-200 μm), but suggests a higher "plasticity" compared to polycrystalline SiC and Al_2O_3.

The data from Table I and new data on the original SiC materials tested by Lundberg and Lundberg [15] (with the exception of SiC-B (L)) are collected in Table III. The SiC samples from Table I are listed as SiC (R), while the Lundberg samples are listed as SiC (L). The slope c is changed to a "plasticity" parameter by taking the absolute value of the reciprocal of c. The last column adds the calculated Knoop hardness at 1N to the new plasticity value in a way that corresponds to what Lundberg et al.[3] did to relate the transitional velocities to compressive yield strength and plasticity.

These results suggest that a plasticity value calculated in the manner described above may be a useful plasticity index/metric and that adding this value to a consistent hardness value (here using the value at 1 N) may be a useful predictor for transitional velocities as measured in the Lundberg et al.[3] approach. The plasticity is seen to vary from the most brittle boron carbide materials - 6.51, 6.67 to AlON -17.70 and SiC-HPN(L) -19.16. However, when the hardness is added to these values it results in a parameter that seems to correspond more closely with the transitional velocities of the SiC samples tested by Lundberg and Lundberg[15].

One very interesting additional issue arose during this work. One of the differences between SiC-N and SiC-B is that SiC-B has been more variable. It became very clear when trying to find correlations to the Lundberg and Lundberg[15] results that Knoop hardness measurements had to made on these same Lundberg samples. The apparent variability of SiC-B clearly made these correlations difficult to impossible when not using the identical Lundberg and Lundberg materials.

Table III. Data from Wilantewicz et al.[14] and new data for the Lundberg and Lundberg[15] SiC samples using Eq. 3 are compiled together with transitional velocities on the latter. A new plasticity parameter and the sum of HK_{1N} and plasticity are also listed.

Material	HK (GPa) Load = 1N	Trans. Vel.(m/s)	Slope (c)	-(1/c) Plasticity	HK_{1N} + plasticity
SiC-N (R)	24.1	-	-0.0699	14.31	38.41
SiC-B (R)	23.8	-	-0.0728	13.74	37.54
PS-SiC	23.8	-	-0.0845	11.83	35.63
AlON	16.0	-	-0.0565	17.70	33.70
AD995 CAP3	17.9	-	-0.0832	12.02	29.92
Cercom B_4C	29.6	-	-0.1536	6.51	36.11
Ceradyne B_4C	28.1	-	-0.1499	6.67	34.77
SiC-N (L)	22.4	1507	-0.0661	15.13	37.53
SiC-B (L)	NA	1549	-	-	-
SiC-SC-1RN (L)	22.9	1526	-0.0659	15.18	38.08
SiC-HPN (L)	22.5	1625	-0.0522	19.16	41.66

CONCLUSIONS

The Vickers hardness-load data of these materials is well described by equations predicting a plateau value of hardness e.g., Eq. 5. Reasons for the nearly constant hardness at high loads are not fully understood, but may possibly be related to the more severe cracking typically seen around Vickers indentations compared to Knoop indentations[7]. The magnitude of the ISE, given by the 'a' values in Table II, is greatest for the pressureless-sintered SiC and less for the hot-pressed SiC's, consistent with the slopes from the Knoop analysis.

The Knoop hardness-load data of these ceramics is well described by a power-law curve ($H = kF^c$) fit. A plot of $\log_{10}$ (HK) vs. $\log_{10}$ (F) yielded easily comparable straight lines, whose slope and intercept data might be useful parameters to characterize the materials. It is proposed that the absolute value of the reciprocal of the slope is a measure of plasticity and that the sum of this value with the calculated Knoop hardness at 1 N may be a useful parameter to correlate to the Lundberg et al.[3] transitional velocity. Since reliable hardness indentations can be attained at higher loads with Knoop compared to Vickers indenters, Knoop testing is better suited for this type of analysis.

Finally, these are preliminary results and clearly should be followed up in more detailed investigations. In addition, since the data was obtained using quasistatic hardness measurements, extrapolation to dynamic environments may not be totally valid and would need to be confirmed by additional systematic work.

REFERENCES

[1]M. Flinders, D. Ray, A. Anderson, R.A. Cutler, "High-Toughness Silicon Carbide as Armor," *J. Am.*
Ceram. Soc., **88** [8] 2217-2226 (2005).
[2]M. Chheda, M.J. Normandia, and J. Shih, "Improving Ceramic Armor Performance," *Ceramic Industry* 124-126 January 2006.
[3] P. Lundberg, R. RenstroKm, B. Lundberg, Impact of metallic projectiles on ceramic targets- transition between interface defeat and penetration: International Journal of Impact Engineering 24 (2000) 259-275.
[4] I. M. Pickup and A.K. Barker, "Damage Kinetics in Silicon Carbide", in Shock Compression of Condensed Matter – 1997, Schmidt, Dandekar and Forbes, eds., pp. 513-516.
[5] B. J. James, "Factors affecting ballistic efficiency tests and the performance of modern armor systems", presented at the European Fighting Vehicle Symposium, Shrivenham, UK, May 1996.
[6] J. W. McCauley and George Quinn, "Special Workshop: Kolsky/Split Hopkinson Pressure Bar Testing of Ceramics", ARL-SR-144, Sept 2006.
[7]J.B. Quinn and G.D. Quinn, "Indentation Brittleness of Ceramics: A Fresh Approach," *J. Mat. Sci.*, **32** 4331-4346 (1997).
[8]Z. Li, A. Ghosh, A.S. Kobayashi, and R.C. Bradt, "Indentation Fracture Toughness of Sintered Silicon Carbide in the Palmqvist Crack Regime," *J. Am. Ceram. Soc.*, **72** [6] 904-911 (1989).
[9]T. Taniguchi, M. Akaishi, and S. Yamaoka, "Mechanical Properties of Polycrystalline Translucent Cubic Boron Nitride as Characterized by the Vickers Indentation Method," *J. Am. Ceram. Soc.*, **79** [2] 547-549(1996).
[10]R. Berriche and R.T. Holt, "Effect of Load on the Hardness of Hot Isostatically Pressed Silicon Nitride," *J. Am. Ceram. Soc.*, **76** [6] 1602-1604 (1993).
[11]J.J. Swab, "Recommendations for Determining the Hardness of Armor Ceramics," *Int. J. Appl. Ceram. Technol.*, **1** [3] 219-225 (2004).
[12]ASTM Standard C 1326-03, "Standard Test Method for Knoop Indentation Hardness of Advanced Ceramics," ASTM International. West Conshohocken, PA, USA 2003.
[13]ASTM Standard C 1327-03, "Standard Test Method for Vickers Indentation Hardness of Advanced Ceramics," ASTM International, West Conshohocken, PA, USA 2003
[14]T.E. Wilantewicz, W.R. Cannon, and G.D. Quinn, "The Indentation Size Effect (ISE) for Knoop Hardness in Five Ceramic Materials," pp. 237-249 in *Advances in Ceramic Armor II, Ceramic Engineering and Science Proceedings, Cocoa Beach, Volume 27, Issue 7*. Ed's. Andrew Wereszczak, Edgar Lara-Curzio, and Lisa Prokurat-Franks (2006).
[15] P. Lundberg and B. Lundberg, Transition between interface defeat and penetration for tungsten projectiles and four silicon carbide materials, *International Journal of Impact Engineering*, 31, 781–792 (2005).

Author Index